Description Approaches and Automated Generalization Algorithms for Groups of Map Objects

Haowen Yan

Description Approaches and Automated Generalization Algorithms for Groups of Map Objects

Springer

Haowen Yan
Faculty of Geomatics
Lanzhou Jiaotong University
Lanzhou, Gansu, China

ISBN 978-981-13-3677-5 ISBN 978-981-13-3678-2 (eBook)
https://doi.org/10.1007/978-981-13-3678-2

Library of Congress Control Number: 2018964703

This Springer imprint is published by the registered company Springer Nature Singapore Pte Ltd.
The registered company address is: 152 Beach Road, #21-01/04 Gateway East, Singapore 189721, Singapore

Preface

Map generalization seeks to maintain and improve the legibility of a map after the scale has been changed. Automated map generalization is a necessary technique for the construction of multi-scale vector map databases that are crucial components in spatial data infrastructure of cities, provinces, and countries. Nevertheless, this is an unrealized dream yet, because many algorithms for map feature generalization are not truly automatic and therefore need human's interference.

In recent decades, scholars in the communities of cartography and geographic information sciences have been making great effort on the theories and methods for automated map generalization and have got many achievements, including various indices, operators, and algorithms for map feature generalization and softwares for automated map generalization (though they are generally semiautomated).

As far as the algorithms for map feature generalization are concerned, the ones for individual objects generalization have been mature and can be put into practical use, while the ones for groups of objects generalization are still on the way and need more investigations, though much progress has been made.

This book emphasizes on representation and generalization of groups of map features. It is well known that map objects can be classified into three categories according to their geometric character, i.e., point, linear, and areal. If only groups of objects on maps are considered, we have point clusters, linear networks, and areal groups. In the meanwhile, if map feature type is considered, we have point clusters, groups of contours, road networks, river networks, continuous areal features, and discrete areal features. Thus, the chapters of the book are arranged in this classification. This book believes that representation methods are as important as generalization algorithms in map generalization; hence, it spends much words on describing map feature representation methods.

The author would like to express his appreciations to many people who made the completion of this book possible. Above all, he is grateful to Dr. Robert Weibel in the Department of Geography, University of Zurich, Switzerland; Dr. Zhilin Li in the Department of Land Surveying and Geo-Informatics, Hong Kong Polytechnic University, Hong Kong; and Professor Jiayao Wang in the PLA Information

Engineering University, China, who discussed many topics with the author years ago at the early stage of this book. Second, the authors feel so indebted to Dr. Xiaomin Lu, Dr. Haiying Wang, Dr. Jianglei Jin, and Dr. Hongyuan Fan who drew graphics and/or checked language for the draft of the book. Last but not least, the author appreciates the Natural Science Foundation of China (No. 41671447 and No. 71563025) and the National Key R&D Program of China (No. 2017YFB0504203) for their financial support to the work described in the book.

The book can be a reference to the graduates and researchers who are interested in cartography and geographic information science/systems, especially those in automated map generalization and spatial database construction. Any comments and suggestions regarding this book are greatly welcomed and appreciated.

Lanzhou, China — Haowen Yan
July 30, 2018

Contents

Chapter 1
Introduction

1.1 Multi-scale Representation

Geometric objects on maps are representations of real phenomena in the geographical space. A single phenomenon may have multiple representations reflecting its different perspectives at the same or different scales on maps. Usually, an object or a group of objects in the geographic space has two types of "multiple representations" on maps. The former is that the object or the group of objects is expressed by different cartographers using different symbols at the same scale on maps, i.e. multi-cartographers' representations or horizontal representation. The latter is that the object or the group of objects is observed at different distances and therefore expressed at different scales on maps, i.e. multi-scale representations or perpendicular representations.

An example of the two types of multiple representations is shown in Figs. 1.1, 1.2 and 1.3. Two cartographers/observers depict the pond (Fig. 1.1) using different polygons, though they observe the pond at the same scale (Fig. 1.2). On the other hand, the pond can also be observed at different distances or heights, and therefore be described using different polygons at different scales (Fig. 1.3).

This book does not care multi-cartographers' representations, but pays attention to multi-scale representations of geographic phenomena on maps. Here, the scale of a map is the ratio of a distance on the map to the corresponding distance on the ground.

Multi-scale representation is characterized by the fact that geographic information may be represented on maps at different levels of detail or generalization. The functionality of multi-scale representations is supported in current geographic information systems (i.e. GIS) by simply storing a series of maps each of which is at a predefined scale. This approach is called "multi-version spatial databases" (MVSD) which can provide direct, simple and efficient solutions for many GIS applications. Nevertheless, MVSD has a number of potential problems such as data inconsistency among different maps, high storage and update overhead of the databases, and

H. Yan, *Description Approaches and Automated Generalization Algorithms for Groups of Map Objects*, https://doi.org/10.1007/978-981-13-3678-2_1

Fig. 1.1 A pond

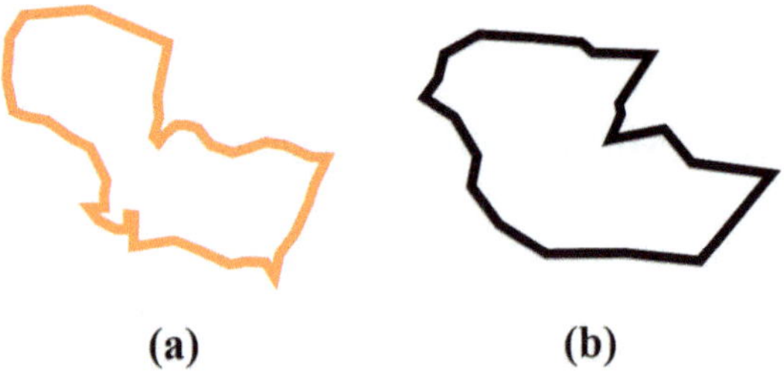

Fig. 1.2 Multiple representations of the pond at the same scale by different cartographers

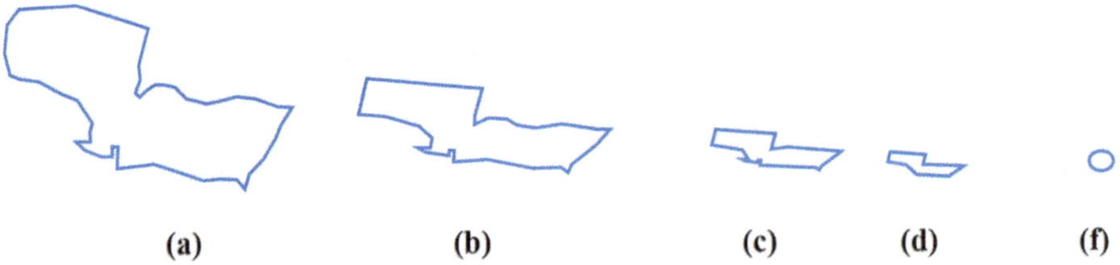

Fig. 1.3 Multiple representations of the pond at four different scales

inflexibility in control over the update of the multi-scale databases. To overcome these shortcomings, many geographers and cartographers have proposed an ideal way: as far as a given region is concerned, only a map database of the region at a larger scale is constructed; the map databases of the region at the other smaller scales can be generated using the larger scale database. This is called "one-version spatial databases" (OVSD). OVSD can ensure all of the representations are consistent, and the amount of map data is not too large, and update of the databases is easy to carry out. Currently, OVSD is the predominant method over MVSD in constructing multi-scale vector map databases, supporting on-demand and on-the-fly mapping, and high quality spatial data retrieval. Hence, this book focuses on automated implementation of multi-scale vector map databases by OVSD, i.e. automated map generalization.

1.2 Digital Earth: Applications of Multi-scale Representation

A vision for a Digital Earth was displayed by then USA Vice President Al Gore in a famous speech (1998): "imagine, for example, a young child going to a Digital Earth exhibit at a local museum... she sees Earth as it appears from space. Using a data glove, she zooms in, using higher and higher levels of resolution, to see continents, then regions, countries, cities, and finally individual houses, trees, and other natural and man-made objects." Here, the description of the Digital Earth presents a multi-scale representation of the physical Earth we reside, and gives us a kind of typical applications of multi-scale maps: what the child sees are the continents, regions, countries, cities, individual houses, trees, and other natural and man-made objects, but that really take effects are the multi-scale spatial databases, and usually each database is at a specific scale.

Google Earth, launched in 2005, is the first successful example of the implementation of the concept of Digital Earth foreseen and put forwards by Al Gore (1998). The product was so innovative that it was proclaimed as the beginning of "geospatial revolution" and the first generation of virtual globe (Geospatial Revolution Project 2010). Since then, a number of digital Earths, digital countries, cyber cities, virtual towns and communities have appeared all over the world. The other similar platforms (e.g. http://map.baidu.com/) have also been built and put into use in many countries. The key theories and techniques that support the implementation of the applications include computational science, mass storage, broadband networks, interoperability, metadata etc. More importantly, besides imagery data at multiple resolutions, maps at multiple scales are also necessary for constructing these platforms and supporting these applications. For example, the China's Fundamental Geographic Information System is one of the typical platforms, and a series of vector map databases at scales of 1:4,000,000, 1:1,000,000, 1:250,000, 1:50,000 and 1:10,000 are important components of the platform. Although the first generation of Google Earth achieves not all of the elements of the Gore vision, the technical achievements of this first generation of virtual globes stimulate the general public becomes more and more engaged with technology and applications of the Digital Earth, and advances in understanding the Earth system and constructing Digital Earths have been made, especially in the construction and application of multi-scale map databases.

1.3 Automated Map Generalization: Implementation of Multi-scale Vector Map Databases

Scale reduction from source maps to target maps inevitably leads to conflict and congestion of map symbols. To make the maps legible, the map features should be deleted, simplified and/or generalized. In the community of cartography this process is named map generalization. From 1960s, automatic generation of the maps at multiple smaller scales using the maps at a larger scale, i.e. automated map generalization, has become a dream of cartographers and geographers. The objective of automated map generalization is to build a multi-scale vector map database that facilitates retrieval of selected types of map data at an appropriate level of detail and renders the data as a legible map.

Automation of map generalization has been viewed as one of the most difficult research issues in the community of cartography and geographic information sciences for decades. There have at least the following three reasons that this difficult problem has not been solved yet by far:

Firstly, features and reliefs in the 3-dimensional geographic space are diverse and complicated; thus it is considerably difficult to mathematically describe and express them in 2-dimensional map spaces.

Secondly, automated map generalization is a simulation of traditional manual map generalization; nevertheless, the resulting maps are usually different for different cartographers even if the same original smaller scale maps are provided for generalization. Thus, it is not easy to describe the generalization procedure of a specific map feature/relief using a solitary algorithm.

Thirdly, automated map generalization is a process depending not only on cartography and geographic information science, but also on psychology and human recognition; however, researchers in these fields cannot explain what cartographers think and how they generalize map features/relief step by step.

Automated map generalization has aroused the interests of many cartographers, geographers, and even scholars in computer science, psychology, graphics mathematics and artificial intelligence in recent decades, and a lot of achievements have been made in the following aspects: (1) concepts and fundamental theories of map generalization; (2) algorithms and operators for map generalization; (3) approaches for controlling map generalization procedures and evaluating map generalization result; and (4) development of software and systems for map generalization engineering. The algorithms are the "protagonist" in implementing automated map generalization, and therefore the achievement on them is much richer than that on other aspects. Nevertheless, current achievements cannot support the construction of a fully automatic map generalization system, yet. This is why MVSD but not OVSD has been used in constructing vector map databases for supporting existing digital Earths, countries, cities, towns etc.

1.4 Review of the Approaches to Describing Map Features and the Algorithms for Automated Map Generalization

As far as automated map generalization is concerned, the following three viewpoints are credible and acceptable by cartographers:

Firstly, algorithms are the core of automated map generalization. If appropriate algorithms can be found for all types of map features and reliefs, it would be not difficult to design and implement an automated map generalization system/software.

Secondly, description of map features and reliefs is the foundation and prerequisite of map generalization algorithm development, and it is as important as design of map generalization algorithms. No good description to a type of map features, no good algorithm can be achieved.

Thirdly, multi-scale map databases are usually used in two kinds of environments, i.e. personal computers and computer networks. No matter what kind of environments the map data are generalized, the essence, fundamental problems and research issues do not change. Therefore, this book does not differentiate the environments that the map data are used.

For the above reasons, the approaches for describing map features and the algorithms for automated map generalization are critically reviewed in the following paragraphs.

1.4.1 Approaches to Describing Map Features

Map features are reflections of landscapes and objects in the geographic space; however, the landscapes and objects are diverse, complicated and/or complex. Therefore, it is a difficult issue to quantitatively describe map features using mathematical formulae, though it is necessary to do so for the purpose of automated generalization of map features in digital environments. Indeed, some research work has been done regarding this issue in recent decades, and the achievements are largely based on knowledge and rules, theories of spatial relations, geometric factors and parameters, and information theory.

1.4.1.1 Approaches Based on Knowledge and Rules

Descriptive knowledge and rules used in map generalization are of great importance and have been explored by cartographers for decades (Barrault et al. 2001; Galanda and Weibel 2002) and they can be classified into four categories (Guo 1998).

1. Knowledge caused by the changes of map scale, i.e. map scale changes and precision changes.
2. Knowledge containing in map features: this includes geometric and semantic characteristics of individual map features and spatial and non-spatial relations between/among map features.
3. Knowledge for guiding the use of map generalization algorithms and operators: it mainly refers to the rules and/or requirements in calling map generalization algorithms, e.g. how to assign values for the parameters of the algorithms, how

to arrange the order of a group of algorithms and operators in a map generalization project, etc.

4. Knowledge for assessing map generalization results: it includes the methods for evaluating the map generalization results of individual algorithms as well as the rules and approaches for evaluating the map generalization quality of a map generalization system/software.

1.4.1.2 Approaches Based on Spatial Relations

Spatial relation is a common factor used in describing map features (Müller and Wang 1992; Regnauld 2001; Li and Huang 2002; Duchêne et al. 2003; Mustiere 2005). Guo (1997) classified spatial relations into five categories, i.e. distance relation, topological relation, direction relation, similarity relation, and correlational relation.

1. Spatial distance relation: it is used for describing the distance between/among spatial objects, and it is the most fundamental one in the five types of spatial relations, and is usually a key factor in calculation of other spatial relations. A distance can be an absolute, a relative, a qualitative, or a quantitative distance.
2. Spatial topological relation: it is one type of the most widely used spatial relations. For the purpose of easy calculation, Guo (1997) classified topological relations into 19 types, e.g. overlap, touch, intersection, point-in-polygon etc.
3. Spatial direction relation: it is used to describe orientation relations between/among spatial objects, usually using the terms such as north, south, west, east, up, down, right, left etc. or quantitative angles such as 56.5°. Qualitative description and quantitative calculation of direction relations have been explored by a number of scholars (Goyal 2000; Yan et al. 2006).
4. Spatial similarity relation: it is a factor for evaluating how similar two spatial objects or two spatial scenes (a group of spatial objects is viewed as a scene) are. Spatial similarity relation is of great useful because multi-scale representation of spatial objects is obviously a kind of similarity transformation in map spaces. Quantitative spatial similarity relations can be used to automate map generalization algorithms and map generalization systems (Yan and Li 2014).
5. Spatial correlative relation: achievements in correlative relations are considerably rich. A typical and useful example of spatial correlative relation is the First Law of Geography and the Second Law of Geography. According to Waldo Tobler's (1970) First Law of Geography: everything is related to everything else, but near things are more related than distant things. This first law is the foundation of the fundamental concepts of spatial dependence and spatial autocorrelation and is utilized specifically for the inverse distance weighting method for spatial interpolation and to support the regionalized variable theory for kriging (Karen 2008). The Second Law of Geography is less well known, which complements the first: the phenomenon external to an area of interest affects what goes on inside.

1.4.1.3 Approaches Based on Geometric Factors

It is an intuitive thought to describe map features using geometric factors. For example, local density, global density and distribution range are used to describe the geometric characteristics of point clusters on maps; road density, road length and area of mesh are used to describe the geometric characteristics of road networks.

1.4.1.4 Approaches Based on Information Theory

Information theory was firstly introduced in cartography by Shkhov (1967, 1970); then it has been developed by a number of scholars (Neumann 1994; Papadias and Sellis 1994; BjǾrke 1996), and four types of information on maps are addressed and discussed by Yan and Wang (2005).

1. Positional or statistical information: it refers to the number of map features, e.g. the number of points in a cluster or a region, the number of settlements in a village.
2. Topological information: it is used to describe topological relations between/among map objects.
3. Thematic information: this type of information is usually expressed using so-called "attributes" in geographic information systems, e.g. area and population are two attributes of cities that are often adopted in spatial databases.
4. Metric information: it is the information of map features and relief described by geometric parameters, e.g. distance between map objects, area of a regions, length of a road etc.

1.4.2 Algorithms for Automated Map Generalization

Features on maps can be classified into three categories, i.e. point, linear and areal features, and the distribution of each category of features on maps is either in individual or clustering/grouping form; thus, there are six types of features on maps, i.e. (1) individual point objects, (2) individual linear objects, (3) individual areal objects, (4) point clustering objects, (5) linear network objects, and (6) areal group objects. Accordingly, there are six types of map generalization algorithms (Table 1.1).

1.4.2.1 Algorithms for Generalizing Individual Point Objects

Individual point objects are the geometrically simplest ones on maps, and they cannot be simplified anymore, and they can be either 'deleted' or 'retained' in map

Table 1.1 Types of features and corresponding map generalization algorithms

Types of features	Example of features	Map generalization algorithms
Individual point object	A control point, an isolated historic object	None
Individual linear object	A linear river, a road, a borderline	The Douglas-Peucker algorithm (Douglas and Peucker 1973); the Visvalingam algorithm (Visvalingam and Williamson 1995); the Mustiere algorithm (Mustiere 2005)
Individual areal object	A areal pond or lake, a polygonal settlement, a patch of agricultural land	The area-patch generalization algorithm (Müller and Wang 1992)
Point clustering objects	Control points in a region, islands containing in an archipelago, settlements in a village	The settlement-spacing ratio algorithm (Langan and Poicker 1986). The circle-growth algorithm (Van Kreveld et al. 1995), the on-line algorithm for point clustering thematic feature simplification (Burghardt et al. 2004), the dot map simplification algorithm (De Berg et al. 2004), the Voronoi-based algorithm (Yan and Weibel 2008)
Linear network objects	Linear rivers in a basin, roads in a district	The stroke-based algorithm for road network generalization (Thomson and Richardson 1999),
Areal group objects	Areal lakes in a political region, rectangular settlements in a neighbourhood	The building displacement algorithm; the building typification algorithms (Regnauld 2001)

generalization. Hence, none of other special algorithms have been developed for this type of objects.

1.4.2.2 Algorithms for Generalizing Individual Linear Objects

Algorithms for generalizing linear features are most popularly used in map generalization operations, because statistically about 80% features on maps are linear ones. The objective of individual linear feature generalization is to reduce the redundant points on the line on larger scale maps so that the linear feature becomes simpler on smaller scale maps, and in the meanwhile the generalized linear feature should be similar to original its counterpart (i.e. the linear feature before generalization) as far as possible.

A number of algorithms have been proposed for simplifying individual linear features, e.g. the Douglas-Peucker Algorithm (Douglas and Peucker 1973), the Reumann-Witkam Algorithm, the Strip-Tree Algorithm (Buttenfield 1986), the BLG-based Algorithm (Zhan and Mark 1993; van Oosterom and Schenkelaars

1995; Saalfeld 1999), the Li-Openshaw Algorithm (Li and Openshaw 1992) etc. These algorithms simplify individual linear features in light of angles between neighbouring line segments, or the lengths of radius vectors (Douglas and Peucker 1973), or the areas of triangles formed by linking neighbouring line segments of the linear feature (Li and Openshaw 1992), or a combination of them (van Oosterom and Schenkelaars 1995).

To sum up, existing algorithms for generalizing individual linear features simplify lines well and can meet the demands of developing automated map generalization systems, and therefore almost no achievements have been made regarding this issue in recent two decades.

1.4.2.3 Algorithms for Generalizing Individual Areal Objects

Settlements, islands, water bodies, and land patches are typical areal objects on large and intermediate scale maps. An individual areal object on the map can be deleted or simplified in the process of map generalization. Whether the areal object is retained or deleted generally depends on the area of the object and its importance compared with the surrounding objects. There are three types of algorithms for simplifying individual areal features: vector-based (Zhang and Tulip 1990; Müller and Wang 1992; Boffet and Serra 2001; Regnauld 2001; Christophe and Ruas 2002; Rainsford and Mackness 2002; Li et al. 2004), raster-based (Monmonier 1983; Su and Li 1995; Su et al. 1997) and integrated algorithms (Li and Su 1996).

1.4.2.4 Algorithms for Generalizing Point Clusters

Control points are typical point objects. In addition, settlements, islands, ponds, wells, and lakes on intermediate and small scale maps are distributed in point clustering form (Fig. 1.4). The objective of point clustering object generalization is to reduce the number of the point objects but keep similarity between the point clusters before and after generalization as far as possible.

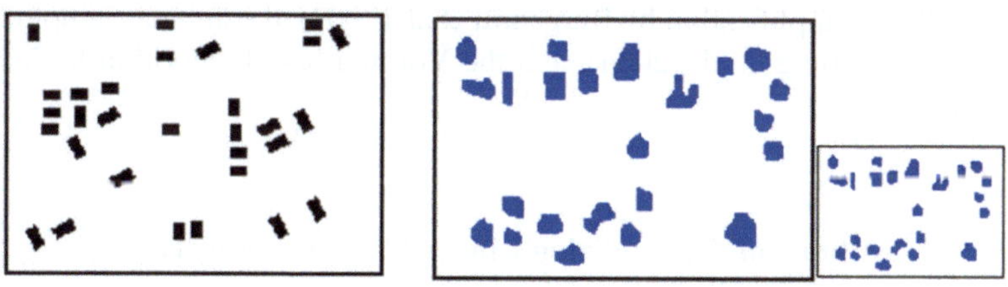

(a) (b)

Fig. 1.4 Point clusters on maps: (**a**) settlements in a small village; and (**b**) polygonal islands on the larger scale map can be viewed as point clusters on the smaller scale map

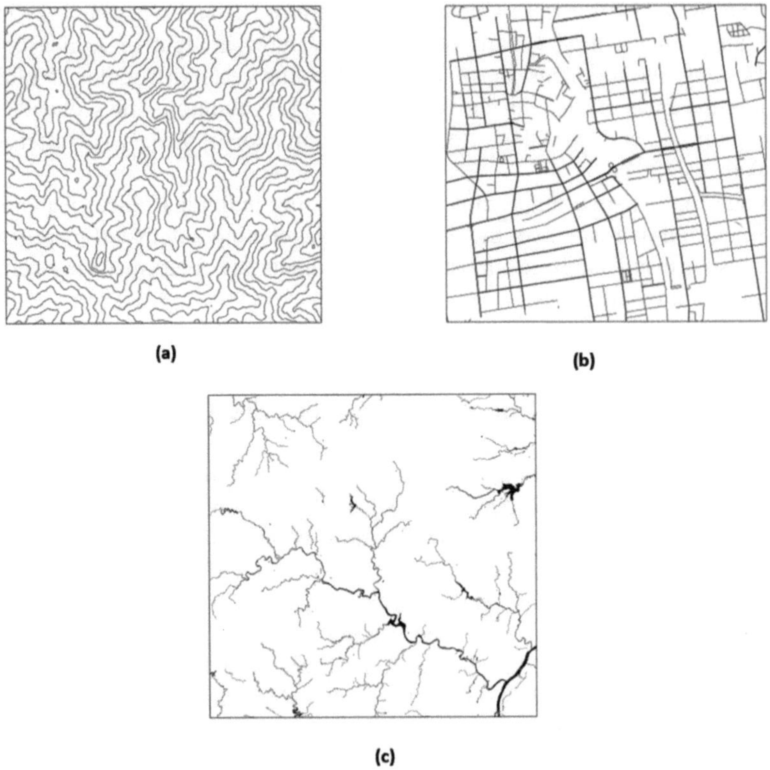

Fig. 1.5 Three types of linear clusters/networks on topographic maps: (**a**) contour line clusters, (**b**) road networks, and (**c**) river networks

Five algorithms for point clustering feature generalization were proposed by Langan and Poicker (1986), i.e. the settlement-spacing ratio algorithm, the distribution coefficient algorithm, the gravity-modeling algorithm, the set-segmentation algorithm, and the quadrat-reduction algorithm. After this, a couple of algorithm have also been proposed, including the circle-growth algorithm for settlement selection by van Kreveld et al. (1995), the on-line algorithm for point clustering thematic feature simplification by Burghardt et al. (2004), the dot map simplification algorithm by De Berg et al. (2004), and the Voronoi-based algorithm by Yan and Weibel (2008).

1.4.2.5 Algorithms for Generalizing Linear Clusters/Networks

There are three types of linear clustering/networks on topographic maps, i.e. contour clusters, road networks, and river networks (Fig. 1.5). However, they are different in spatial distribution, spatial semantics, graphics, and geometric structure; thus, they

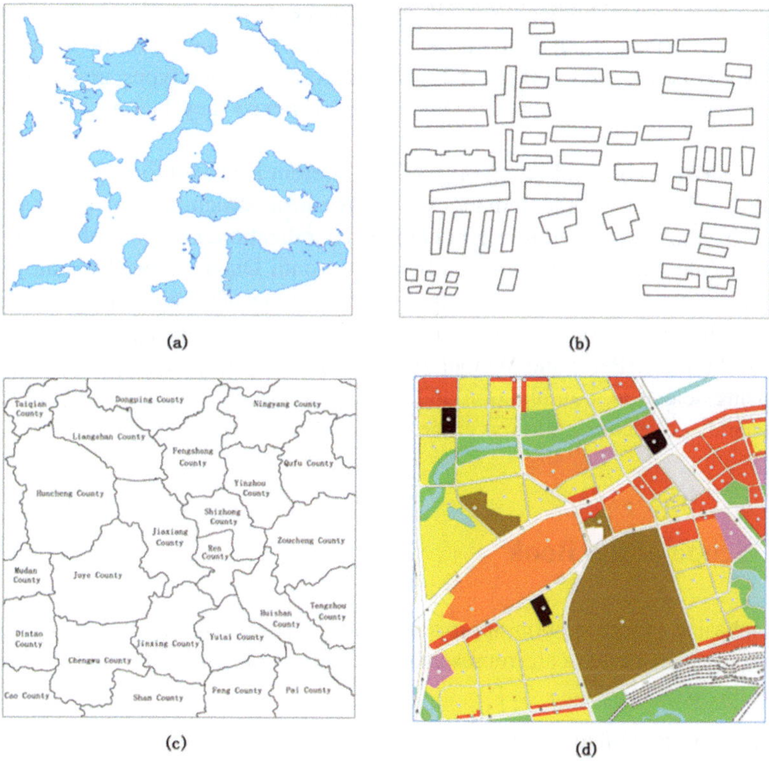

Fig. 1.6 Four types of areal group/clustering objects on maps: (**a**) water bodies, (**b**) settlements, (**c**) political regions, and (**d**) land types using different colors representing different types of lands

are described using different parameters and/or approaches and generalized by developing different operators/algorithms.

Actually, algorithms have been developed for different type of linear feature networks/clusters on maps. For example, contour clusters can be generalized by considering digital terrain models and minor valley braches (Ai and Li 2010) or multiply represented using contour trees (Guilbert 2013); road networks are simplified by taking into account blocks (Gülgen and Gökgöz 2011) or using artificial neural networks (Zhou and Li 2014); and river networks can be generalized using Fractal Geometry, Horton's laws and Tokunaga cyclicity (Tarboton 1996).

1.4.2.6 Algorithms for Generalizing Areal Group Objects

Water bodies, settlements, political regions, and land types on large scale maps are generally polygonal objects (Fig. 1.6). However, they are topologically different; therefore, they can be classified into two categories according to topological relations between polygons: discrete areal groups, and connected areal groups. It is obvious that water bodies as well as settlements are topologically separated (Fig. 1.6a, b); hence both of them belong to the first category. While the polygons representing political regions as well as that representing land types are topologically adjacent, i.e. they are connected (Fig. 1.6a, b); hence they belong to the second category.

The four types of areal group objects are different in spatial topological relations, spatial distribution, spatial semantics and geometric structure, thus different algorithms have been designed for them in map generalization. For example, settlements on large scale maps are generalized by the Voronoi Diagram and Delauney Triangulation (Li et al. 2004), or by taking into account Gestalt theory (Regnauld 2001); types of land patches on categorical maps are generalization using vector and raster-based techniques (Peter and Weibel 1999); groups of lakes and ponds are generalized using selection, combination and aggregation operators; and political regions are generalized according to the grades of political regions.

1.5 Scope of the Book

To emphasize on description approaches and generalization algorithms of types of map features in multi-scale representations, this book limits its topics within the following scopes.

- This study is limited within two-dimensional map spaces.
- The main spatial data in this book are vector map data.
- We pay more attention to topographic map data in the book.
- The algorithms addressed in the book should be automatic.
- We do not distinguish between personal computers and the Internet as far as the environment of the algorithms is concerned.
- Contexts of the generalized map features are generally not taken into consideration in the algorithms.

1.6 Organization of the Book

Topographic map features can be classified into six categories in light of Table 1.1: individual point objects, individual linear objects, individual areal objects, point clustering objects, linear network objects, and areal group objects. On the other hand, linear network objects include contour line groups, road networks and river networks, and areal group objects include two types of discrete areal objects (water bodies and buildings) and two types of connected areal objects (land patches and political regions). Hence, there are totally 11 types of topographic map features need to be considered in map generalization: (1) individual point objects, (2) individual linear objects, (3) individual areal objects, (4) point clustering objects, (5) contour line groups, (6) road networks, (7) river networks, (8) water bodies, (9) settlements or buildings, (10) land patches, and (11) political regions.

In light of the critical review in Sect. 1.4.2, generalization of individual point, linear and areal features has basically solved; thus, this study emphasizes on the description approaches and generalization algorithms of groups/clusters/networks of map features, and it is organized as follows: after this chapter (Introduction), description approaches and generalization algorithms for the other eight types of map group/network/cluster features will be addressed in detailed chapter by chapter (Chaps. 2, 3, 4, 5, 6, 7 and 8).

It should be emphasized that potential research issues regarding corresponding type of map features will be presented at the end of each chapter. In addition, some concluding remarks and discussions will be given in Chap. 8.

References

Ai T.H., Li J.Z., 2010. A DEM generalization by minor valley branch detection and grid filling, ISPRS Journal of Photogrammetry and Remote Sensing, 65(2): 198–207

Barrault M. Regnauld N. Duchene C. et al. 2001. Integrating multi-agent, object-oriented and algorithmic techniques for improved automated map generalization. In Proceedings of the 20th International Cartographic Conference, Beijing, China, pp. 2110–2116.

BjØrke J. 1996. Framework for entroy-based map evaluation. Cartography and Geographic Information Systems, 23(2), pp.78–95.

Boffet A., Serra S.R., 2001. Identification of spatial structures within urban blocks for town characterization, the Proceedings of the 20th International Cartographic Conference, Beijing, China.

Burghardt D., Purves R., and Edwards A., 2004, Techniques for on-the-fly generalization of thematic point data using hierarchical data structures. In the Proceedings of the GIS Research UK 12th Annual Conference, Norwich, UK.

Buttenfield B.P., 1986. Digital definitions of scale-dependent structure, Auto-Carto1986, Vol.1: 497–506.

Christophe S., Ruas A., 2002. Detecting building alignment for generalization purpose. The Proceedings of the 10th International Symposium on Spatial Data Handling, Ottawa, Canada.

De Berg M., Bose P., Cheong O. and Morin P., 2004, On simplifying dot maps. Computational Geometry, 27(1): 43–62.

Douglas D.H., Peucker T.K., 1973. Algorithms for the reduction of the number of points required to represent a line or its caricature, The Canadian Cartographer, 10(2): 112–122.

Duchêne C. Bard S. and Barillot X. 2003. Quantitative and qualitative description of building orientation. in The 5th ICA workshop on progress in automated map generalization, Paris, France.

Galanda M., and Weibel R. 2002. An agent-based framework for polygonal subdivision generalization. In Proceedings of Spatial Data Handling 2002, Ottawa, Canada. (CD-ROM).

Geospatial Revolution Project., 2010, The Pennsylvania State University. Retrieved from: http://geospatialrevolution.psu.edu/ Accessed: March 8, 2018.

Gore A, 1998, The Digital Earth: Understanding our planet in the 21st Century. Al Gore's speech at California Science Center, Los Angeles, California, on January 31, 1998. Retrieved from: http://www.isde5.org/al_gore_speech.htm Accessed: December 24, 2016.

Goyal R K. 2000. Similarity assessment for cardinal directions between extended spatial objects. PhD thesis, The University of Maine

Guilbert E., 2013. Multi-level representation of terrain features on a contour map, Geoinformatica, 17(2): 301–324.

Guo Q.S., 1998, Classification and formal description of the knowledge in automated map generalization, Journal of the Surveying and Mapping University of PLA, 15(3): 199–203.

Guo R.Z., 1997. Spatial analysis. Wuhan: Press of Wuhan University of Science and Technology. (in Chinese)

Gülgen F., Gökgöz T., 2011. A block-based selection method for road network generalization, International Journal of Digital Earth, 4(2): 133–153.

Karen K. 2008. Encyclopedia of Geographic Information Science, SAGE, p146–147

Langan C., and Poicker T., 1986, Integration of name selection and name placement. in the Proceedings of the 2nd International Symposium on Spatial Data Handling, pp. 50–64.

Li Z.L., Huang P.Z., 2002. Quantitative measures for spatial information of maps. International Journal of Geographical Information Systems, 16(7): 699–709.

Li Z.L., Openshaw S., 1992. Algorithms for objective generalization of line features based on the natural principle. International Journal of Geographical Information Systems, 6(5): 373–389.

Li Z.L., Su B., 1996. Algebraic models for feature displacement in the generalization of digital map data using morphological techniques, Cartographica, 32(3): 39–56.

Li Z.L., Yan H.W., Ai T.H., Chen J., 2004. Automated building generalization based on urban morphology and gestalt theory. International Journal of Geographical Information Science, 18 (5):513–534.

Monmonier M., 1983. Raster-mode area generalization for land use and land cover maps, Cartographica, 20(4): 65–91.

Müller J C., Wang Z.S., 1992. Area-patch Generalisation: A Competitive Approach, The Cartographic Journal, 29(2), pp.137–144.

Mustiere S., 2005. Cartographic generalization of road in a local and adaptive approach: a knowledge acquisition problem. International Journal of Geographical Information Science, 19(8–9): 937–956.

Neumann J. 1994. The topological information content of a map: an attempt at a rehabilitation of information theory in cartography. Cartographica, 31(1): 26–34.

Papadias D. and Sellis T. 1994. The qualitative representation of spatial knowledge in two dimensional space. Very Large Database Journal, 3(4): 479–516.

Peter B., Weibel R., 1999. Using vector and raster-based techniques in categorical map generalization, the Third ICA Workshop on Progress in Automated Map Generalization, Ottawa, 12–14 August, 1999.

Rainsford D., Mackness W., 2002. Template matching in support of generalization of rural buildings. The Proceedings of 10th International Symposium on Spatial Data Handling. Ottawa, Canada.

Regnauld N., 2001. Contextual building typification in automated map generalization. Algorithmica, 30(2): 312–333.

Saalfeld A., 1999. Topologically consistent line simplification with the Douglas-Peucker Algorithm, CGIS 1999, 26(1): 7–18.

Su B., Li Z.L., 1995. An algebraic basis for digital generalization of area-patches based on morphological techniques, The Cartographic Journal, 32(2): 148–153.

Su B., Li Z.L., Lodwick G., Muller J.C., 1997. Algebraic models for the aggregation of area features based on morphological operators, International Journal of Geographical Information Systems, 11(3): 233–246.

Sukhov V., 1967. Information capacity of map entropy. Geodesy and Aerophotography, 10(4): 212–215.

Sukhov V., 1970. Application of information theory in generalization of map contents. International Yearbook of Cartography, Vol.10: 41–47.

Tarboton D.G., 1996. Fractal river networks, Horton's laws and Tokunaga cyclicity, Journal of Hydrology, 187: 105–117.

Thomson R.C., Richardson D.E., 1999. The "good continuation" principles of perceptual organization applied to the generalization of road networks. The Proceedings of ICA 1999, Ottawa, Canada.

Tobler W.R., 1970. A computer movie simulating urban growth in the Detroit region. Economic Geography, 46: 234–240.

Van Kreveld M., Van Oostrum R., Snoeyink J., 1995. Efficient settlement selection for interactive display. The Proceedings of Auto Carto 12, Bethesda, Md. pp.287–296.

van Oosterom P., Schenkelaars V., 1995. The development of an interactive multi-scale GIS, International Journal of Geographical Information Systems, 9(5): 489–507

Visvalingam M., Williamson P.J, 1995. Simplification and generalization of large scale data for roads, Cartography and Geographic Information Science, 22(4), 3–15.

Yan H.W., Chu Y.D., Li Z.L., Guo R.Z., 2006. A quantitative description model for directional relations based on direction groups, Geoinformatica, 10(2):177–195.

Yan H.W., Li J., 2014. Spatial similarity relations in multi-scale map spaces. Springer International Publishing Switzerland.

Yan H.W., Wang J.Y., 2005, A Voronoi-based generic algorithm for point cluster generalization, Journal of Image and Graphics,10(5): 633–636.

Yan H.W., Weibel R., 2008. An algorithm for point cluster generalization based on the Voronoi diagram, Computers & GeoSciences. 34(8): 939–954

Zhan F., Mark D.M., 1993. Conflict resolution in map generalization: a cognitive study, Auto-Carto 1993, Vol.13: 406–413.

Zhang G., Tulip J., 1990. An algorithm for the avoidance of sliver polygons and clusters of points in spatial overlay. The Proceedings of the 4th International Symposium on Spatial Data Handling, Zurich, Switerland.

Zhou Q., Li Z.L., 2014. Use of artificial neural networks for selective omission in updating road networks, The Cartographic Journal, 51(1): 38–51.

Chapter 2
Description and Generalization of Point Clustering Features

This chapter aims at presenting the algorithms for point clustering feature generalization. For this purpose, it firstly defines and describes the relevant concepts (Sect. 2.1) and illustrates the types of point clustering features on maps (Sect. 2.2), and analyzes the approaches for describing point clustering features (Sect. 2.3). After this, it presents and analyzes the existing algorithms (Sects. 2.4 and 2.5). Last, the chapter is ended by a concluding summary (Sect. 2.6).

2.1 Multi-scale Representations of Point Clustering Features

Point features refer to the objects showing on the map using point symbols. The objects might be physical entities (e.g., ponds, buildings, patches of agricultural land, temples, tombs, etc.) in the geographic space, and they also might be economic or human phenomena. When they are shown on the map at a small scale, they are usually expressed using point symbols such as dots, triangles, rectangles, crossed, etc.

As far as a specific type of point features is concerned, although the symbols are physically separated on the map, they are, in essence, logically correlated and can be viewed as clusters in people's spatial recognition according to their spatial relations such as distance and topological relations. We may call them point clustering features. When the map scale become smaller, the point symbols usually become crowded and the map becomes illegible. Hence, those unimportant features need to be deleted from the map. The less the map scale becomes, the more the points should be deleted. Generalization of areal/point clustering features can generate a series of maps at multiple scales which comprise a multi-scale representation (e.g. Fig. 2.1).

At least two types of point clustering features can be discriminated on maps considering the original geometric characteristics of the map symbols.

H. Yan, *Description Approaches and Automated Generalization Algorithms for Groups of Map Objects*, https://doi.org/10.1007/978-981-13-3678-2_2

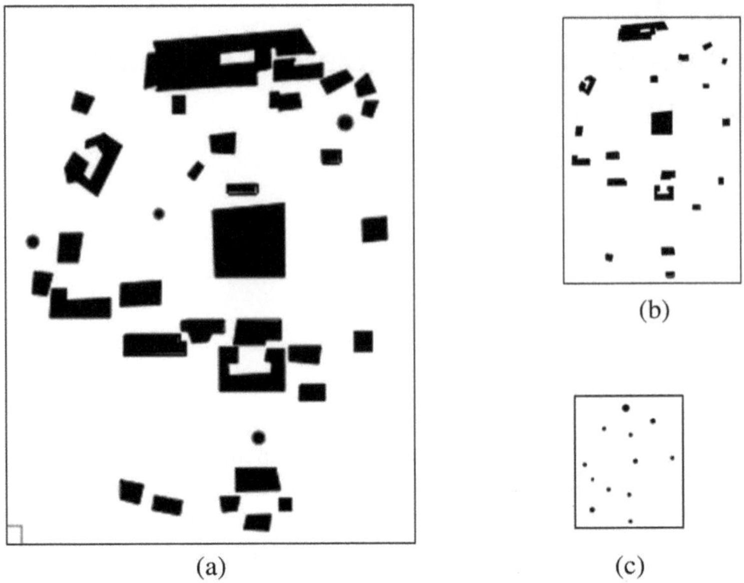

Fig. 2.1 A multi-scale representation of the point clustering settlements (the maps are not always shown to the exact scales). (**a**) at scale 1:10K; (**b**) at scale 1:25K; and (**c**) at scale 1:50K

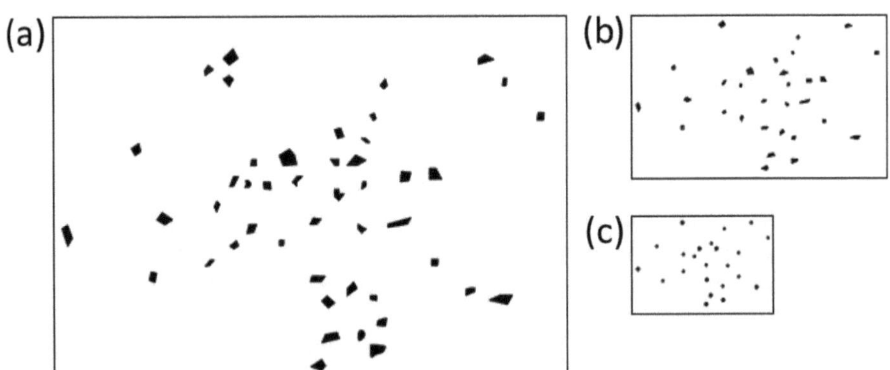

Fig. 2.2 A multi-scale representation of the settlements consisting of areal and point symbols. (**a**) Settlement clusters consist of areal and point symbols at scale 1:10K. (**b**) Settlement clusters consist of areal and point symbols at scale 1:25K. (**c**) Settlement clusters consist of areal and point symbols at scale 1:50K

1. Mixture of areal and point features

The original features consist of areal symbols and point symbols (e.g. the settlements in Fig. 2.2a). When they are generalized the areal symbols become smaller and smaller (Fig. 2.2b), and gradually all of they have to be expressed using point

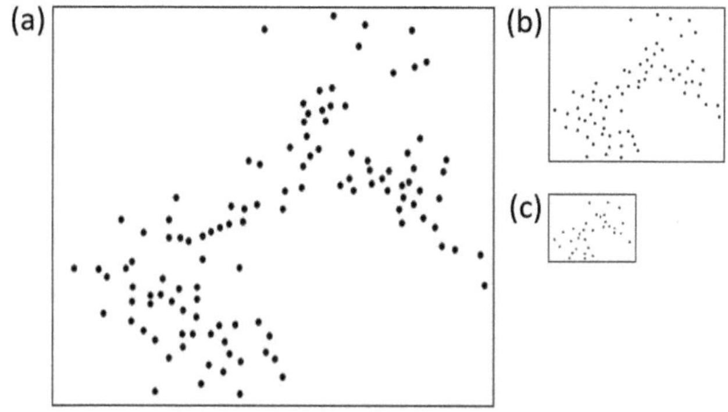

Fig. 2.3 A multi-scale representation of gas stations consisting of point symbols on the original map. (**a**) Gas stations on the map at scale 1:5K. (**b**) Gas stations on the map at scale 1:10K. (**c**) Gas stations on the map at scale 1:25K

symbols (Fig. 2.2c). Typical examples of this type include settlements in rural areas in China and archipelagos or islands in the Pacific Ocean and the India Ocean.

2. Pure point features

In most cases, point clustering features on maps consist of only point symbols on maps (Fig. 2.3a). Generalization of such point clustering features results in a series of maps at multiple scales that comprise a multi-scale representation of the point clustering features (Fig. 2.3a–c).

2.2 Types of Point Clustering Features on Maps

A number of types of map features can be viewed as point clusters on maps.

- Control points: There are usually many classes of control points (Fig. 2.4) which have very high positioning accuracy and are used as the "skeleton" of the surveyed area. They form point clusters on the map, and they are generally regarded as a whole in map generalization.
- Settlements: Settlements on small scale maps (scale $S < 1:100K$) are usually very tiny, point-like symbols, and the settlements in a specific areas can be viewed as a combination of a number of point clusters (e.g. Fig. 2.5).
- Islands: Islands on small scale maps (the scale is less than 1 : 100K) are small polygons, so they can be regarded as point clustering features in map data processing (e.g. Fig. 2.6).
- Isolated small-size features: Trees, traffic lights, post boxes, temples, towers, gas stations, and even schools and hospitals are shown on small scale maps (the scale is less than 1 : 50K) using point symbols. When a specific type of these features is processed, the features in the studied area are generally regarded as point clusters (Figs. 2.7 and 2.8).

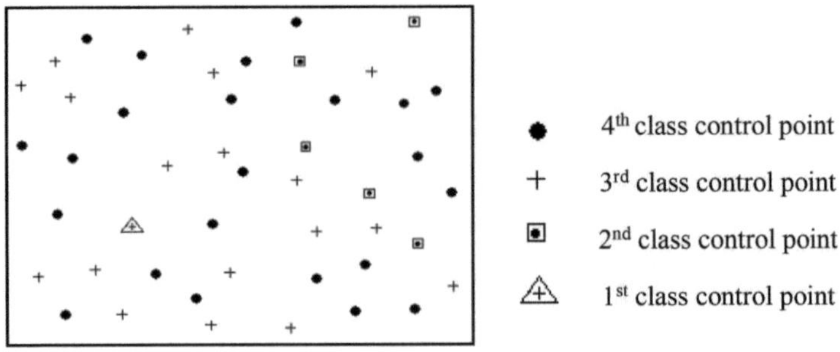

Fig. 2.4 Control points are point clustering features on maps

Fig. 2.5 Settlements in rural areas of south China on small scale maps form point clusters

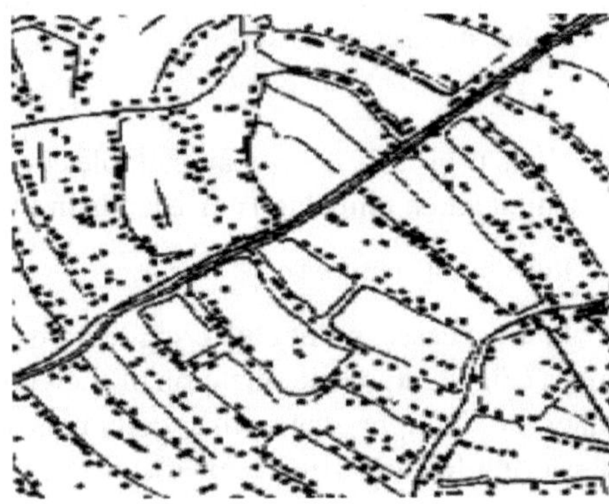

Fig. 2.6 Islands on small scale maps can be viewed as point clusters

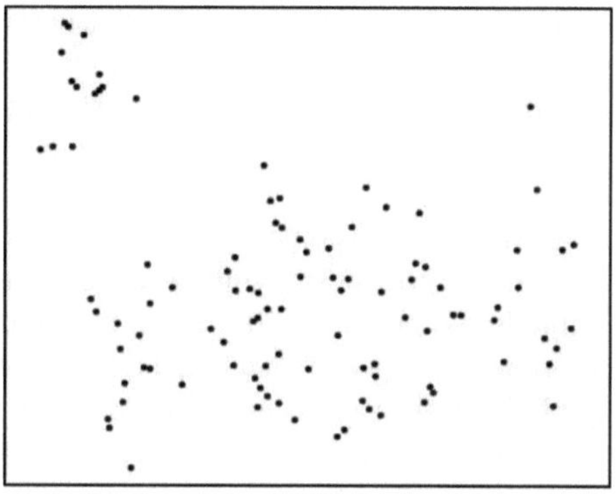

Fig. 2.7 Distribution of supermarkets in a district of Guangzhou, China

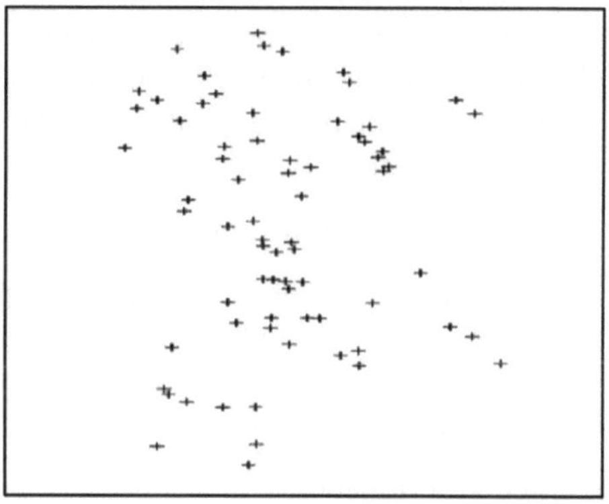

Fig. 2.8 Distribution of hospitals in a district of Beijing, China

2.3 Approached to Describing Point Clustering Features

It is necessary to know how to describe point clustering features on maps before designing and developing algorithms to generalize them. Literature shows that the cartographic information theory Sukhov (1967, 1970) and geometric measures (Ahuja 1982; Yan and Weibel 2008) are two commonly used approaches.

2.3.1 Information Contained in Point Clusters

The concept of information was firstly used in the communication theory by Shannon and Weaver (1949), and they proposed so-called 'Entropy' which is a quantitative measure for the information content contained in a message. After years, this concept was introduced to the cartographic community by Sukhov (1967, 1970) and four types information had been proposed ever since.

The four types of information discussed below can be generally considered in all types of map features, thus they should be naturally useful in evaluating the information containing in point clustering features on maps.

- Statistical information

Sukhov (1967) considered the statistics of different types of symbols on the map. The entropy of these symbols is computed using the proportion of each type of symbols to the total number of the symbols as the probability. Such a type of information is purely statistical information and the spatial distribution of the symbols on the map is not considered, as is pointed out by Li and Huang (2002).

Bjørke (1996) considered three types of information, i.e. positional, metric, and topological. The positional information of a map considers all the occurrences of the map features as unique events and all map events are equally probable. That is, the positional entropy is simply computed by counting the number of map features. It is obvious that the positional information discussed by Bjørke (1996) is the same as the statistical information by Sukhov (1967, 1970), so we use the term statistical information in this book.

- Topological information

Neumann (1994) introduced the concept of topological information by considering the connectivity and adjacency between map features. The topological information considers different types of relations between map features such as overlap, intersection, containing and contained etc.

- Metric information

The metric information proposed by Bjørke (1996) considers the variation of the distance between map features, while Li and Huang (2002) uses the size of the area occupied by the Voronoi diagram of each map feature, instead of the distances between map features.

- Thematic information

The thematic information considers categorical difference of neighboring map features (Neumann 1994), and it can also be used to discriminate the quantitative differences of the same type map features (Yan and Weibel 2008).

2.3.2 *Measures for Types of Information Contained in Point Clusters*

Previous research (Ahuja 1982; Ahuja and Tuceryan 1989; Guo 1997; Sadahiro 1997; Yan and Weibel 2008) has shown that a number of measures can be used for the description of point clusters.

- The number of the points: It refers to the number of the points in the studied region (e.g. Fig. 2.9a).
- Importance value (Van Kreveld et al. 1997; Langran and Poicker 1986): It is a value assigned to a point as a measure of its importance among other points. For example, it is either 1 or 2 in Fig. 2.9b.
- Neighboring points: They are those points that share adjacency relations with a given point (e.g. Fig. 2.9c). They may be the Voronoi neighbors, the fixed radius neighbors, k-nearest neighbors (Ahuja 1982; Ahuja and Tuceryan 1989), etc.
- Distribution range: It is usually a polygonal area that contains all studied points (Guo 1997) (e.g. Fig. 2.9d).
- Absolute local density: This means the average number of points in a unit area, or average distance between points (Sadahiro 1997).
- Relative local density: It is a ratio of absolute local density at a certain location to the summation of the absolute local density over the whole studied region (Sadahiro 1997).
- Distribution modes (Guo 1997): This refers to one or more areas with much higher relative local density compared with their surroundings (e.g. Fig. 2.9e).
- Distribution axes: It refers to one or more axes extracted from the point clusters whose extension is linear (Guo 1997) (e.g. Fig. 2.9f).

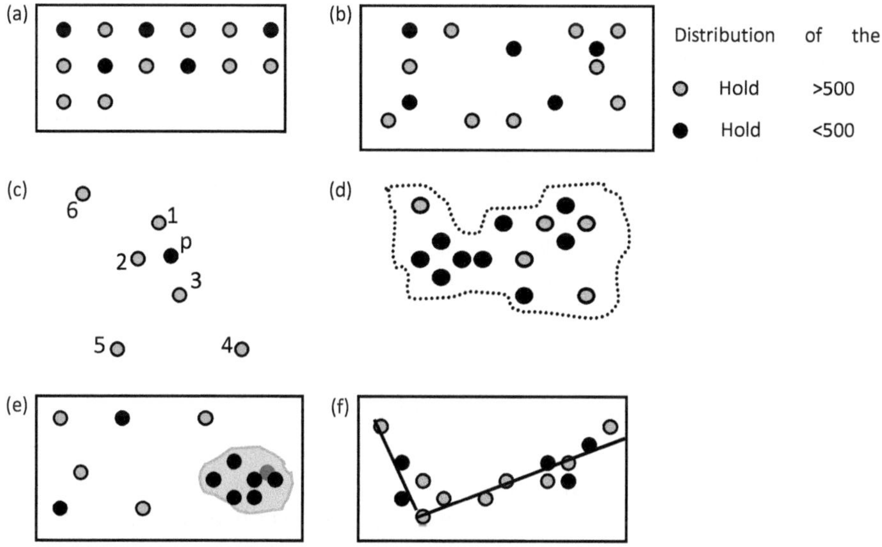

Fig. 2.9 Measures for describing point clusters. (**a**) The number of points: there are 14 points in the cluster. (**b**) Weights: the weight of each black point is 1, while that of the grey point is 2. (**c**) Neighbours: points 1, 2 and 3 are the neighbours of P. (**d**) Distribution range: the area enclosed by the dotted line. (**e**) Distribution centre: the grey area. (**f**) Distribution axis: the line crossing the points

Table 2.1 Corresponding relations between the measures and the four types of information containing in point clustering features

Types of information	Measures
Statistical	The number of the points
Thematic	Importance value
Topological	Neighbouring points (Voronoi neighbours, fixed radius neighbours, k-nearest neighbours, etc.)
Metric	Absolutely local density, relative local density, distribution range, distribution mode, distribution axis, etc.

Each of the above measures corresponds to one of the four types of information discussed in Sect. 2.3.1. Their relations can be summarized and listed (Table 2.1). From Table 1, it is clear that statistical information and thematic information can only be described using one factor respectively, while topological information and metric information possess multiple measures.

2.4 Algorithms for Point Clustering Features Generalization

A main objective of point clustering feature generalization is to reduce the number of the point objects and in the meanwhile keep similarity of the point clusters before and after generalization in distribution range, distribution density, distribution mode, neighbouring relations etc. A number of algorithms for point clustering feature generalization have been proposed by far. This includes six algorithms for settlement selection proposed by Langran and Poicker (1986) and van Kreveld et al. (1995), the online algorithm for point clustering thematic feature simplification (Burghardt et al. 2004; Weibel and Burgardt 2008), the dot map simplification algorithm by De Berg et al. (2004) and the Voronoi-based algorithm by Yan and Weibel (2008).

2.4.1 Algorithms for Settlement Selection

Langran and Poicker (1986) proposed five algorithms for settlement selection, including the settlement-spacing ratio algorithm, the distribution coefficient algorithm, the gravity-modeling algorithm, the set-segmentation algorithm, and the quadrat-reduction algorithm, and following them van Kreveld et al. (1995) proposed the circle-growth algorithm. Because the set-segmentation algorithm and the quadrat-reduction algorithm require much human intervention in settlement selection and are not suitable for automated map generalization, thus this section excludes them but discuss the other four algorithms.

2.4.1.1 The Settlement-Spacing Ratio Algorithm

The idea of the settlement-spacing-ratio algorithm is as follows: supposing that a circle around each settlement is placed around each settlement whose size is inversely proportional to its importance value, the radius R of each circle can be calculated by:

$$R = c/i \tag{2.1}$$

Where, c is a constant, and i is the importance value of the settlement.

Then the settlements are accepted in the order of their importance values, starting with the one with the greatest importance value. A settlement may only be accepted if its circle contains none of the previously accepted settlements. In other words, a settlement with small importance value can only be accepted if it is isolated from other ones. Here, the constant of proportionality c determines how many settlements are accepted. It is obvious that smaller values for c mean smaller circles and this generally leads to more settlements being selected for display.

This method doesn't have the monotonicity property and that a complete ranking of the settlements cannot be calculated. In fact, it can be that no value of c gives a desired number of settlements. It is also possible that two different selections have the same size. This is all caused by the fact the monotonicity property is not respected by the model.

2.4.1.2 The Distribution Coefficient Algorithm

The distribution-coefficient algorithm works in light of so-called "nearest neighbor index". The nearest neighbor index is the ratio of the actual mean distance to the nearest neighbor and the expected mean distance to the nearest neighbor. At beginning, all settlements are arranged in decreasing order of their importance values. Starting with a small set of largest ones, a settlement may only be accepted if its addition to the already accepted set does not decrease the nearest neighbor index. Although the number of settlements in the final selection is fixed, it may be changed by introducing a tuning factor.

This algorithm has two advantages. Firstly, it does not result in a complete ranking of the settlements; and secondly, the actual importance value of a settlement is only used in the order of processing, not in the selection.

2.4.1.3 The Gravity-Modeling Algorithm

This algorithm is based on a concept called "the influence of settlement". The influence of one settlement on another one can be calculated by dividing the importance value of the first one (the selected one) by the distance to the other one. At beginning, all settlements are arranged in decreasing order of their importance value. A settlement s can only be accepted if its importance value is greater than the summed influence of all already selected settlements on s. Similar to the settlement spacing-ratio algorithm, a constant of proportionality c is also used. The next settlement under consideration may be accepted if the summed influence of the already accepted settlements on the candidate is less than c times the importance value of the candidate. The number of selected settlements can be adjusted by changing c.

This model, like the spacing-ratio algorithm, does not respect the monotonicity property and therefore does not give a complete ranking of the settlements.

2.4.1.4 The Circle-Growth Algorithm

The circle-growth algorithm gives a ranking of the settlements: supposing that a circle is drawn for each settlement, the radius of the circle is proportional to the importance value of the settlement, similar to Formula 2.1. The initial constant of proportionality c should be determined to ensure that no two circles are overlapped. Then c is increase, causing all circles to grow, until the circle of a settlement fully covers that of some other one. The former circle is viewed as dominating the latter one, and the latter has the lowest rank of all settlements and is removed. This process can be repeated while assigning higher and higher ranks, until only the most important settlement remains.

This algorithm meets the following two important requirements that need to be satisfied in map generalization:

1. When two settlements compete for space on the map, the more important one can survive.
2. A settlement with less importance value can be retained on the map if there are no settlements with higher importance values in its proximity.

The drawback of this algorithm is that a settlement with considerably great importance value can have a global effect on the map which means its neighborhood occupies a very large area of the map, and its neighbouring settlements are suppressed. On the contrary, if there is large region where there is no settlement of great importance value, a great number of the settlements might be selected.

To overcome the above shortcomings, three variations of the circle-growth algorithm has been proposed (van Kreveld et al. 1995).

- Variation 1

The key of the issue is the radii of the circles. To control the radii, let the size of the circle be proportional to i^a ($0 \leq a \leq 1$) but not i. By tuning a, the influence of the importance values of the settlements on the selection can be reduced.

- Variation 2

The main drawback of the circle-growth algorithm is that the settlement with a very great importance value has too much influence on the selection than its neighbours, and this may result in the opposite of preserving the local density of the settlement. To overcome this drawback, an improved algorithm is developed for ranking the settlements.

Step 1, the settlement with the greatest importance value is assigned the highest rank.

Step 2, the settlement that is the second in rank is determined by applying the circle-growth idea. The settlement whose circle is covered last by the circle of the settlement with the highest rank is chosen, and its importance value is set to that of its dominating settlement.

Step 3, step 2 is iteratively executed, ranking a next settlement when its circle is covered last by any one of the ranked settlements, until the last settlement is processed.

By this improve algorithm the distribution of the selected settlements can be expected to be more even than the distribution of the selection resulting from the circle-growth algorithm.

- Variation 3

In the previous two improved algorithms, all calculations are done with absolute importance values of the settlements. The third variation makes qualitative rather than quantitative comparisons. Firstly, the settlements are increasingly sorted by their importance values from low to high. Each settlement receives a number that denotes its position in the sorted order. Then this number is used to replace the importance value, after which the ranking is computed using the same method in the circle-growth algorithm. Circles of selected settlements have equal size, and the size of the circles of the not-selected settlements is proportional to their positions in the list sorted on their importance values.

2.4.2 The Dot Map Simplification Algorithm

The term dot map refers to the use of points or dots placing on a map to represent the distribution of a given geographical, human, or societal phenomena. No doubt, dot maps are important and their use and mapping methods have been extensively studied in cartography. There are indeed a number of issues involved in using dots representing spatial distributions. For example, the radii and the color of the dots and/or if the dots are allowed to overlap are important visual considerations. If it is a population dot map, topographic background of the mapped area should be taken into account: a dot map representing population density should not draw dots inside water bodies such as lakes, rivers and seas, and in mountainous areas one altitude has to be taken into account.

Simplification of dot maps has been studied by De Berg et al. (2004) and an algorithm has been proposed by them. This question can be described as follows:

given a point set P_0 representing a distribution on the map at a certain scale s_0. How can we automatically simplify it to get a new representative point set P_1 at a given smaller scale s_1? This question may exist in the following two cases: (1) when people wish scale a dot map. This means the number of the points on the map has to decrease when the size of the map is decreased, otherwise the map would become too crowded; and (2) when people produce the initial dot map. For example, a printed dot map for population distribution should be constructed at a larger scale based on more detailed information such as roads, rivers, lakes, mountains, settlements and houses, and then reduced to the final scale.

No matter which case it is, the essence of the question is "to get an approximate point set P_1 of the original point set P_0." Here, how can we determine the quality of P_1? In other words, an appropriate quantitative measure for evaluating the degree of similarity between $P_0$and P_1 is necessary. De Berg et al. (2004) defined a formal notion called "ε–approximation" as the measure. A set P_1 with m point is called the ε–approximation P_0 of with n points with respect to a family R of ranges, if for any range r ∈ R we have.

$$||r - P_1|/n - |r - P_0|/m| \leq \varepsilon \tag{2.2}$$

Then the approximation error of a set P_1 of m points with respect to a set P_0 of n points (with m ≤ n) for certain families of ranges, namely unit squares, arbitrary squares, and arbitrary rectangles can be obtained, by which the quality of the approximation can be evaluated.

2.4.3 *The On-the-Fly Point Clustering Thematic Feature Generalization Algorithms*

For decades map generalization has focused on the development of automated and assisted techniques for generalizing topographic data (Li et al. 2004; Mustiere 2005; Sester 2005), typically provided by national mapping agencies of many countries. In general, such generalization operations are often pre-computed using traditional operators and stored in multi-scale databases, while on-the-fly generalization denotes the use of automated generalization techniques in real-time. This process creates a temporary, generalized dataset exclusively for visualization, not for storage or other purposes. Thus, on-the-fly generalization is closely connected with highly interactive applications of cartography such as web mapping, mobile mapping (e.g. location-based services), and real-time decision support systems (e.g. disaster and evacuation management) that involve multiple spatial scales. As it takes place in a highly interactive setting, the cartographic quality requirements are typically relaxed compared to traditional, high-quality paper maps (Burghardt et al. 2004; Weibel and Burgardt 2008). By far the algorithms suitable for on-the-fly generalization of topographic data have been developed for application with small devices by Harrie et al. (2002), adaptive zooming by Cecconi and Galanda (2002), and point clustering thematic feature simplification by Burghardt et al. (2004).

Thematic maps traditionally display one, two or no more than a few themes and their variation over space usually with a topographic backdrop. Burghardt et al. (2004) proposed an algorithm for generalizing the thematic maps portraying animal locations. The algorithm deals with thematic data from two different sources: (1) animal locations collected in the Swiss National Park that are delivered to users on handheld mobile devices as part of an information service, and (2) the results of spatial queries to a geographic search engine. In both cases the thematic data are stored as point coordinates with related attributes. Using these data, the following three sets of constraints are identified for delivering visualizations of thematic data to users in real time.

Firstly, the visualizations should be cartographically acceptable, which are intended to maximize legibility on small devices, i.e. minimum symbol size should maintain legibility, and symbols should not be allowed to overlap or touch.

Secondly, any generalization operations should be performed in real-time, i.e. the user's waiting for a response should be minimal; and

Thirdly, visualization of thematic data should maintain meaningful relationships. To maintain meaningful relationships between points, these relationships must first be identified. Techniques which operate only on the point positions and attributes assume some form of second order effect where the position of one point influences another. For instance, in considering the positions of several animals (i.e. deer), hinds may be clustered around a stag displaying such a second order relationship. However, if the deer are separated by some physical feature in the landscape the cluster may have no biological meaning and thus not be relevant. Equally, the deer may be found in areas where a preferred foodstuff occurs and thus variation in density of the deer over space is controlled by some other parameter displaying a first-order effect.

The procedures and principles of the algorithms are described as follows.

The concept of quadtree for partitioning space is used here to allow rapid selection and exaggeration of points at different levels of detail. Moving vertically through the quadtree is directly analogous to zooming operation, whilst moving horizontally between nodes is comparable to panning operation. Each node contains a count of the number of objects existing within the block it represents. The count at the nodes for the current level of zoom is used to generate a visualization using simple selection and exaggeration rules (Fig. 2.10).

Obviously, the quadtree tessellates the space until every point is accounted for by a separated block. When zooming through the data a level is chosen (Fig. 2.10b) that meets a minimum acceptable symbol size criterion.

The quadtree provides an efficient implementation which allows the application of simple rules to generazise thematic point data for panning and zooming whilst meeting basic cartographic constraints. However, points are aggregated according to their quadtree block, with no consideration given to relative adjacency to other points or underlying controls on point distribution.

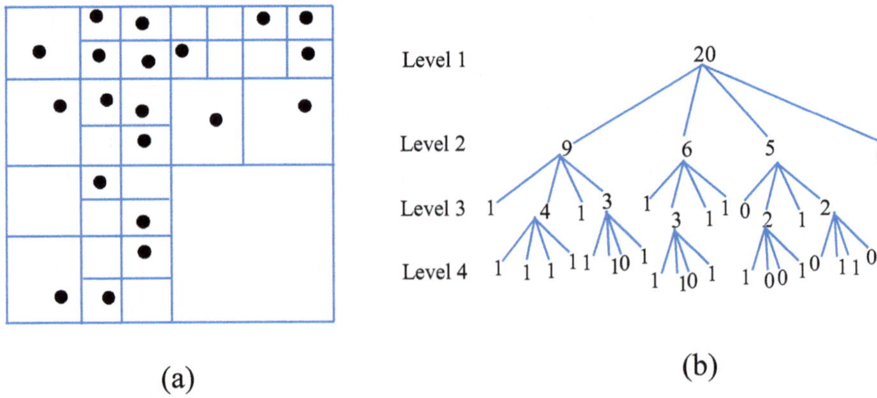

Fig. 2.10 An example of quadtree

2.4.4 The Voronoi-Based Algorithm

The purpose of point cluster generalization is to correctly transmit information from larger scale maps to smaller scale maps, and statistical, thematic, topological and metric information need to be considered in this process. So whether the four types of information are taken into account and transmitted properly should be an important criterion for designing and evaluating algorithms for point cluster generalization. Yan and Weibel (2008) proposed such an algorithm based on Voronoi Diagram, and we call it "the Voronoi-based algorithm" in this chapter.

The algorithms proposed before the Voronoi-based algorithm either do not take into account metric or topological information, or do not use appropriate measures for the description of the four types of information contained in point clustering features. Therefore, the Voronoi-based algorithm was naturally proposed to fill the gap. It takes into account the four types of information, and select the measures are carefully for corresponding types of information (Table 2.1). The reasons why the measures are selected for evaluating a specific type of information are described below in detail.

- For statistical information, "the number of the points" is the only measure and it is used by all the algorithms to control the exact number of points that need to be displayed on the target maps, therefore it should be employed in the Voronoi-based algorithm.
- For thematic information, "importance value" is the only measure, and it is used in the algorithms by Langran and Poicker (1986) and the one by van Kreveld et al. (1997) to evaluate the importance degrees of the settlements so that the settlements with higher importance values can own greater probabilities to be retained on target maps. Hence, "importance value" should be selected in the Voronoi-based algorithm.
- For topological information, the measure called "Voronoi neighbour" is selected.

The settlement–spacing ratio algorithm employs a circle whose radius is inversely proportional to its population (importance value) for each settlement to define adjacency relations between settlements. The distribution-coefficient control algorithm uses the nearest neighbor index as the measure. The circle-growth algorithm uses a circle whose radius is proportional to its population (importance value) for each settlement to define adjacency relations between settlements. So it is clear that three of them use the fixed radius neighbors (in essence, the circles in the settlement–spacing ratio algorithm and the circle-growth algorithm, and the nearest neighbor index in the distribution-coefficient control algorithm are for computing the fixed radius neighbors). However, it has been found that the measure "fixed radius neighbors" has some limitations for describing neighboring relations compared to Voronoi neighbors (Ahuja 1982; Ahuja and Tuceryan 1989).

- For metric information, all the algorithms use only distance as the measure. It is obvious that this measure is too simple to give a good description of density change of point clusters, and it cannot ensure metric information is transmitted correctly. Thus, the Voronoi-based algorithm uses a combination of the four measures, i.e. absolutely local density, relative local density, distribution range and distribution mode.

The methods for integrating these measures into the algorithm are as follows.

1. The number of the points:

The Radical Law or the law of selection (Töpfer and Pillewizer 1966) is employed to determine the number of the points retained on target maps.

$$n_2 = n_1 \sqrt{\frac{S_1}{S_2}} \tag{2.3}$$

Where, n_1 is the number of the point on the source map; n_2 is the number of the point on the resulting map; S_1 is the denominator of the scale of the source map; and S_2 is the denominator of the scale of the resulting map.

2. Importance value

In order to evaluate the change of importance values over the whole region, mean importance value is defined as:

$$\bar{I} = \sum_{i=1}^{n} I_i / n \tag{2.4}$$

Where, $\bar{I}$ is the mean importance value; I_i is the importance value of the ith point; and n is the number of the studied points. The importance value of the ith point is taken into consideration along with the area of its Voronoi polygon, by which the selection probability of the ith point is computed as:

$$P_i = \frac{I_i A_i}{\sum_{k=1}^{n} I_k A_k} \tag{2.5}$$

Where, P_i is the selection probability of the ith point; A_i is the area of the Voronoi polygon of the ith point; I_i and n have the same meanings as that in the Function (2.4).

Here, the numerator $I_i A_i$ is similar to a 'weight' for expressing the importance degree of the ith point: the larger the importance value and the area of the Voronoi polygon of the ith point are, the more likely the point can be retained on the generalized maps (the denominator is a constant, same for all points).

3. Voronoi neighbours

As we discussed in the previous sections of this chapter, the concept of "neighbours" of a point has been defined in many ways: fixed radius neighbors, k-nearest neighbors, Voronoi neighbors, and various extensions to nearest neighbors (Ahuja 1982; Tobler 1970). In the Voronoi-based algorithm, the concept called "k-order Voronoi neighbor" is defined and used. The definition is as follows:

- Point P is the 0-order Voronoi neighbor of itself; and
- If the Voronoi polygon of point Q shares a common edge with that of a (k − 1)-order Voronoi neighbor of P, Q is defined as a k-order Voronoi neighbor of P.

Where, k = 1,2,. . .,n.

Figure 2.11 shows an example for explaining the concept of k-order Voronoi neighbor: as far as point p is concerned, it is the 0–order neighbour of itself; the points annotated 1 is the 1-order neighbours of P, and that annotated 2 is the 2-order neighbours of P, and so on.

The choice of Voronoi neighbours in the Voronoi-based algorithm is guided by the following advantages that the definition of Voronoi neighbours offers over the other neighborhood definitions:

- It is parameter free, while the other definitions have parameters that need to be specified;
- Unlike fixed radius neighbours, it is adaptive to scale and density variations;
- Unlike k-nearest neighbors, the number of neighbours is not fixed; and
- The neighboring relations between any two points are symmetric (see Fig. 2.11: point P and any one of its k-order neighbours are k-order neighbors of each other.

4. Relative local density

This concept is chosen to evaluate the density variations between points before and after generalization. The relative local density of the ith point is defined as:

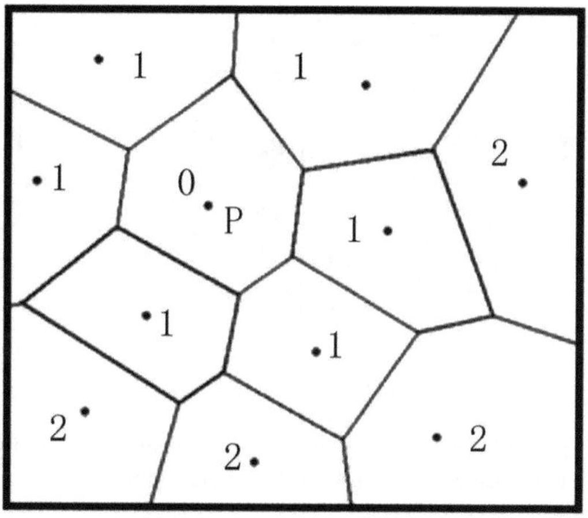

Fig. 2.11 Definition of k-order neighbours

$$r_i = \frac{R_i}{\sum_{k=1}^{n} R_k} \quad (2.6)$$

Where, r_i is the relative local density of the ith point; n is the number of the points in the studied area; R_i is the absolute local density of the ith point and can be calculated by:

$$R_i = 1/A_i \quad (2.7)$$

Where, A_i is the area of the Voronoi polygon containing the ith point.

This definition of absolute local density is a variation of the one given by Sadahiro (1997, p. 52): "a ratio of the local density at the certain location to the summation of local density over the region", while the definition here is the inverse of the area of the Voronoi polygon of the point. The improvement of this definition compared with the one proposed by Sadahiro (1997) is that this one can give absolute (and relative) local density of every point while the former cannot. This makes the comparison of density changes point to point before and after generalization possible.

5. Distribution range

Ahuja and Tuceryan (1989) defined the border of a point cluster as a polygon connected by the hull points. However, the result calculated by this method cannot be directly used in map generalization, because it does not take into account the influence of the points on the border. In other words, the distribution range of the point clusters on maps is generally larger than the one calculated by Ahuja's method (Ahuja and Tuceryan 1989).

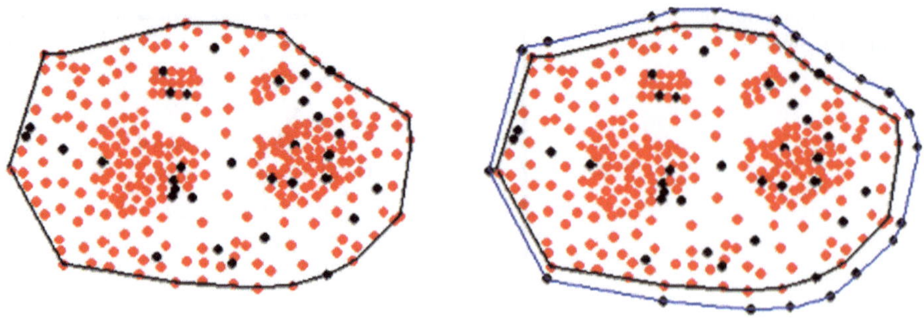

(a) (b)

Fig. 2.12 Methods for describing distribution range of point clusters: (**a**) Ahuja's method; (**b**) the method used in the Voronoi-based algorithm

Taking control points (or supermarkets, or hospitals) as an example (since they are usually displayed on maps using point symbols), each control point can be used as a control in a wide range of area for the purpose of surveying, hence a polygon computed using Ahuja's method (Fig. 2.12a) does not give a good description of the potential 'influence' of these points. Thus, it seems natural to find a larger polygon containing all points and covering the influencing area of the points (Fig. 2.12b).

To construct such polygons, Delaunay triangulation network of the original point clusters is constructed, by which a boundary polygon including all points can be obtained (Fig. 2.12a). Then a pseudo boundary (Fig. 2.12b) is constructed for the original point clusters: each edge of the pseudo boundary is parallel to the corresponding edge of the boundary polygon. The distance between a pseudo point and a corresponding point on the original boundary is equal to the average length of the edges on the original boundary.

The procedure of the Voronoi-based algorithm is an iterative generalization process with the aids of the Voronoi diagram, including the following four steps.

- Step 1, construct the Voronoi diagram for the new point set (i.e. inclusion of psudo-boundary points).
- Step 2, calculate the significance value of each point, and all of the points (except the pseudo points) are sorted in a decreasing order of their significance values.
- Step 3, examine each point (say point P) in the sorted array. Point P is deleted from the point cluster, if the following rules are satisfied: (1) $I_p < R_t I_a$, and (2) none of the points in the adjacent Voronoi polygons of point P has been deleted.

  Where, I_p is the importance value of point p; I_a is the average importance value of all points; R_t is the ratio of selection ratio of all points calculated by the radical law.
- Repeat this step until no more point can be further deleted.

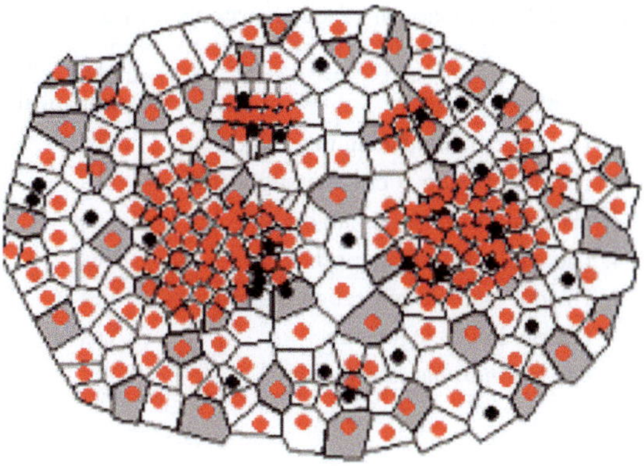

Fig. 2.13 Inner point simplification method (points in shadowed polygons are will-be-deleted)

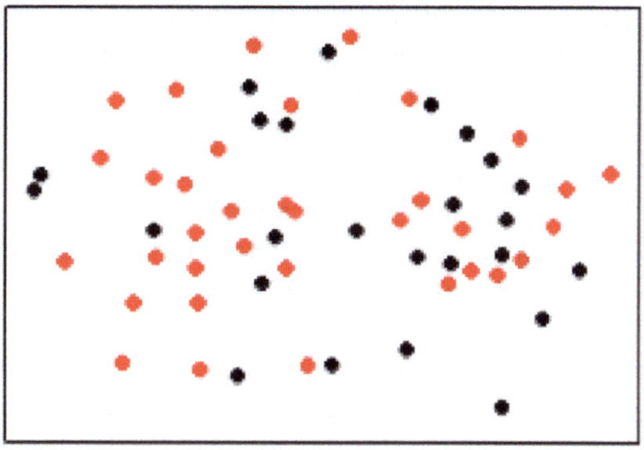

Fig. 2.14 Result generated by the Voronoi-based algorithm using data in Fig. 2.13

- Step 4, if the number of retained point number is less than that calculated by the radical law, end the procedure; or else start a new round of generalization from the first step by regarding the retained points as a new point set (Figs 2.13 and 2.14).

After the last round of deletion, the number of retained points is less than that calculated by the radical law, whereas at the round before the last, the number of retained points must be greater than that calculated by the radical law. This is because the Voronoi-based algorithm deletes a group of points in each round of generalization as they are regarded as being of the same importance. This is different from the circle-growth algorithm that a complete ranking list of point features is given and points are deleted individually.

2.5 Discussions

Purpose of generalizing point clustering features is to reduce the number of the points and in the meanwhile to retain the information containing in original point features. Thus, evaluation of the quality of the algorithms may rely on a comparison

Table 2.2 Comparison of the algorithms for point clustering feature generalization

	Types of information transmitted			
Algorithms	Statistical	Thematic	Topological	Metric
1. The settlement-spacing ratio algorithm	Yes	Yes	No	Yes
2. The distribution coefficient algorithm	Yes	Yes	No	Yes
3. The gravity-modeling algorithm	Yes	Yes	No	Yes
4. The circle-growth algorithm	Yes	Yes	No	Yes
5. The dot map simplification algorithm	Yes	Yes	No	Yes
6. The on-the-fly algorithm	Yes	Yes	No	No
7. The Voronoi-based algorithm	Yes	Yes	Yes	Yes

of the four types of information containing in the original and generalized point clustering features. Table 2.2 presents a comparison of the seven algorithms in transmitting the four types of information containing in the process of point clustering feature simplification.

A number of insights can be gained from this table and our previous sections.

Firstly, only the Voronoi-based algorithm considers all of the four types of information, and uses corresponding parameters to calculate cartographic information in the process of point cluster simplification, while each of the other six ones considers two or three types of information. Especially, they do not take into account topological information among points.

Secondly, the three algorithms (i.e. the settlement-spacing ratio algorithm, the distribution coefficient algorithm, and the gravity-modeling algorithm) proposed by Langran and Poicker (1986) and the circle-growth algorithm proposed by van Kreveld et al. (1995) are all for settlement selection. The main difference among them is that the three ones by Langran and Poicker (1986) do not give a ranking of the settlements in the process of settlement generalization, while the one by van Kreveld et al. (1995) does. After ranking, selection of any number of settlements becomes easy. Moreover, if additional more settlements need to be selected, all previously selected settlements are still retained and do not need to be changed, which is not the case in the three algorithms by Langran and Poicker (1986). In this sense, the circle-growth algorithm is much better than the previous ones.

Thirdly, the dot map simplification algorithm and the on-the-fly algorithm are design for specific maps. The former is only for simplifying dot maps such as dot population maps, and the latter is for generalization and visualization of thematic features that can be viewed as points on maps such as animal in mountainous areas. In other words, both of them are not general algorithms and therefore cannot be commonly in point clustering feature generalization.

Fourthly, each algorithm for point clustering features burdens at least three important tasks: (1) the algorithm is fully automatic; (2) visualization of resulting points is acceptable; and (3) resulting and source points are similar in spatial distribution. The seven algorithms listed in Table 2.2 obviously do not need human's interventions in point feature generalization, and they all have also considered visualization of the points on resulting maps. However, they do not consider how to evaluate the similarity relations between source and resulting point clustering feature except the Voronoi-based algorithm. Not only the Voronoi-based algorithm take into account the four types of information in designing the algorithm, but also the method for comparing each type of information between the resulting map and the source map are given.

Lastly, our future research on this topic may pay more attention to parallel algorithms and online algorithms for point clustering feature generalization in the web age (Jones and Ware 2005).

2.6 Summary of the Chapter

Simplification of a single point is meaningless and worthless to be researched. Thus, this chapter does not care generalization of individual points but focuses on the approaches for simplifying point clustering features on maps when map scale becomes smaller. It presents the types of point clustering features on maps, and addresses the approaches for describing point clustering features which lay a theoretical foundation for the succeeding discussion of the point feature generalization algorithms. The algorithms are the key of this chapter. It gives a sketch of the algorithms and introduces and analyzes the existing seven algorithms in detail. In addition, it discusses the advantages and disadvantages and the potential research issues in point clustering feature generalization.

References

Ahuja, N., 1982. Dot pattern processing using Voronoi neighborhoods. IEEE Transactions on Pattern Analysis and Machine Intelligence 4 (3), 336–343.

Ahuja, N., Tuceryan, M., 1989. Extraction of early perceptual structure in dot patterns: integrating region, boundary and component gestalt. Computer Vision, Graphics and Image Processing 48 (3), 304–356.

Bjørke, J., 1996. Framework for entroy-based map evaluation. Cartography and Geographic Information Systems 23 (2), 78–95.

Burghardt, D., Purves, R., Edwards, A., 2004. Techniques for on the-fly generalization of thematic point data using hierarchical data structures. In: Proceedings of the GIS Research UK 12th Annual Conference, Norwich, UK. http://www.geo.unizh.ch/_burg/literatur/gisruk_draft.pdf.

Cecconi, A., Galanda, M., 2002, Adaptive zooming in web cartography, Computer Graphics Forum, 21, pp. 787–799.

De Berg, M., Bose, P., Cheong, O., Morin, P., 2004. On simplifying dot maps. Computational Geometry 27 (1), 43–62.

Guo, R., 1997. Spatial Analysis, first ed. Press of Wuahan Technical University of Surveying and Mapping, Wuhan, 236pp (in Chinese).

Harrie, L., Sarjakoski, T., Lehto, L., 2002, A variable-scale map for small-display cartography. In Proceedings of the Joint International Symposium on Geospatial Theory, Processing and Applications (Ottawa), CD-ROM.

Jones, C.B., Ware, J.M., 2005. Map generalization in the web age. International Journal of Geographical Information Science 19 (8–9), 859–870.
Langran, C., Poicker, T., 1986. Integration of name selection and name placement. In: Proceedings of second International Symposium on Spatial Data Handling, Washington, USA, pp. 50–64.
Li, Z., Huang, P., 2002. Quantitative measures for spatial information of maps. International Journal of Geographical Information Systems 16 (7), 699–709.
Li, Z., Yan, H., Ai, T., Chen, J., 2004. Automated building generalization based on urban morphology and gestalt theory. International Journal of Geographical Information Science 18 (5), 513–534.
Mustiere, S., 2005. Cartographic generalization of road in a local and adaptive approach: a knowledge acquisition problem. International Journal of Geographical Information Science 19 (8–9), 937–956.
Neumann, J., 1994. The topological information content of a map: an attempt at a rehabilitation of information theory in cartography. Cartographica 31 (1), 26–34.
Sadahiro, Y., 1997. Cluster perception in the distribution of point objects. Cartographica 34 (1), 49–61.
Sester, M., 2005. Optimization approaches for generalization and data abstraction. International Journal of Geographical Information Science 19 (8–9), 871–897.
Shannon, C., Weaver, W., 1949. The Mathematical Theory of Communication. University of Illinois Press, Urbana, IL, 117pp.
Sukhov, V., 1967. Information capacity of a map entropy. Geodesy and Aerophotography 10 (4), 212–215.
Sukhov, V., 1970. Application of information theory in generalization of map contents. International Yearbook of Cartography 10 (1), 41–47.
Töpfer, F., Pillewizer, W., 1966. The principles of selection. The Cartographic Journal 3 (1), 10–16.
Tobler, W.R., 1970. A computer movie simulating urban growth in the Detroit region. Economic Geography 46 (2), 234–240.
Van Kreveld, M., Van Oostrum, R., Snoeyink, J., 1995. Efficient settlement selection for interactive display. In: Proceedings of Auto Carto 12, Bethesda, MD, USA, pp. 287–296.
Van Kreveld, M., Van Oostrum, R., Snoeyink, J., 1997. Efficient settlement selection for interactive display. In: Proceedings of Auto Carto 13, Bethesda, MD, USA, pp. 287–296.
Weibel, R., Burgardt, D., 2008, On-the-Fly generalization. In: Shekhar, [et al.]. Encyclopedia of GIS. New York, US, pp.339–344.
Yan, H., Weibel R., 2008, An algorithm for point cluster generalization based on the Voronoi diagram. Computers & GeoSciences. 34(8): 939–954.

Chapter 3
Description and Generalization of Contour Lines

3.1 Definition and Characteristics of Contour Lines

A contour line (also isoline, isopleth, or isarithm) is a line on a map joining points of equal height or depth above or below a level, usually mean sea level. It is often just called a "contour". Each contour is a closed curve (Veregin 1999, 2000; Cheung and Shi 2004). The difference in height or depth between successive contour lines is the contour interval. It generally becomes greater and greater when the map scale becomes less and less (Joao 1998).

Contour line is a basic technique that is used for representing relief, terrains and landforms on the surface of the Earth (Cromley 1991). A contour map is a map illustrated with contour lines, for example a topographic map (Fig. 3.1), which can show not only the changes of general terrains, but also peaks, valleys, hills, and the steepness or gentleness of slopes using previously specified symbols (Fig. 3.2).

Contour lines have a number of principal characteristics, listed as follows:

(1) All points on a contour line are of the same elevation.
(2) The variation of vertical distance between any two contour lines is assumed to be uniform.
(3) No two contour lines can meet or cross each other except in the rare case of an overhanging vertical cliff or a wall.
(4) A contour line must close itself but need not be necessarily within the limits of the map.
(5) Contours do not pass through permanent structures such as buildings
(6) Contours do not have sharp turnings.
(7) The horizontal distance between any two contour lines indicates the amount of slope and varies inversely on the amount of slope. Thus, contours are spaced equally for uniform slope, closely for steep slope contours and widely for moderate slope.

H. Yan, *Description Approaches and Automated Generalization Algorithms for Groups of Map Objects*, https://doi.org/10.1007/978-981-13-3678-2_3

Fig. 3.1 A topographic map consisting of contour lines

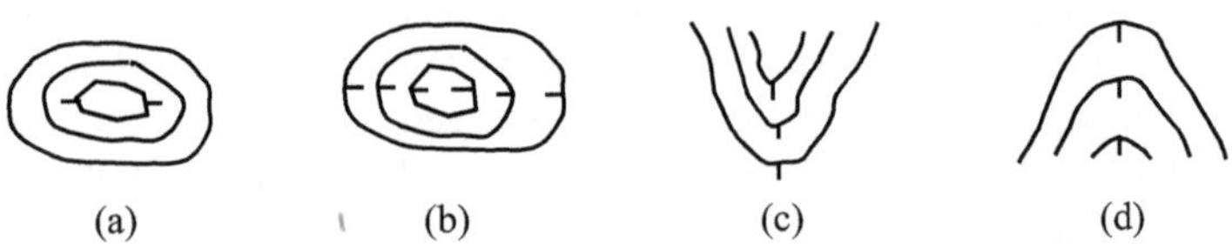

Fig. 3.2 A number of typical terrains expressed using contour lines (**a**) Peak (**b**) Depression (**c**) Ridge (**d**) Valley

(8) A closed contour line on a map represents either depression or hill. A set of ring contours with higher values inside depicts a hill (Fig. 3.2a), whereas the lower value inside depicts a depression (without an outlet) (Fig. 3.2b).

(9) Contour lines of ridge show higher elevation within the loop of the contours, while contour lines cross ridge at right angles (Fig. 3.2c).

(10) Contours deflect uphill at valley lines and downhill at ridge lines. Contour lines in U-shape cross a ridge and in V-shape cross a valley at right angles (Fig. 3.2d). The concavity in contour lines is towards higher ground in the case of ridge and towards lower ground in the case of valley.

3.2 Contour Tree: A Representation of Groups of Contour Lines

Generalization of relief on the topographic map can usually be equal to the generalization of contour line. In the process of generalization, contour lines are simplified, deleted, or moved s, taking into account the basic characteristics of the local

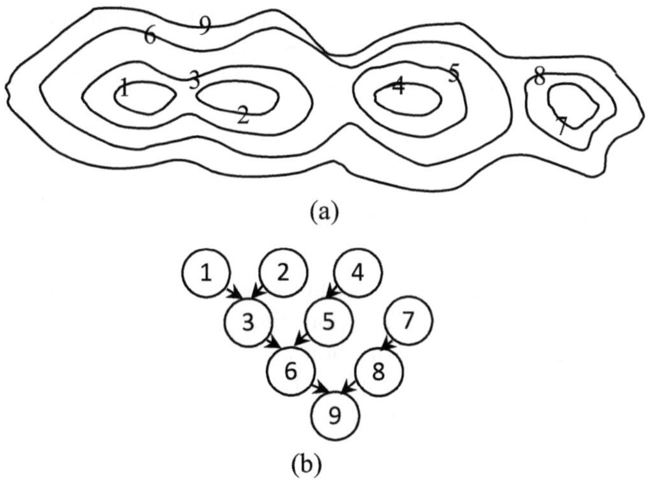

Fig. 3.3 A contour tree

relief and the relations among the relief and other map features, so that the contour lines on the resulting map can be adaptive to the scale and the purpose of the map.

This chapter does not care the approaches to simplifying a single contour line (Jones and Abraham 1987; Wang and Muller 1998), nor the approaches to deleting or/and moving a single contour line; on the contrary, it focuses on how to generalize a group of contour lines, i.e. a number of successive contour lines on the map. Thus, it is natural and important to efficiently express the spatial relations among contour lines.

"Tree" has been used to describe the spatial relation among contours, and it is called contour tree. Contour trees are used to describe the hierarchical relations among adjacent contours, e.g. the parent-child relations between the contours having containment relations and the sibling relations among/between adjacent contours having same elevations. Figure 3.3 shows the principle for constructing contour trees. Here, Fig. 3.3a is a contour map; Fig. 3.3b is the contour tree of the contour map. In the contour tree, contour 1 and contour 2 are siblings; contour 3 is the parent of contour 1 and contour 2.

Contour trees can be used in many fields. For example, contour trees can provide meaningful spatial hierarchical relations when the contour map is used to generate DEM (digital elevation models). In addition, contour trees are used to make three-dimensional topographic maps, because contour trees can easily provide the information of hills, valleys, ridges, saddles, etc. Most importantly, contour trees can be employed in automated generalization of contour maps, because not only may contour trees provide relations between contours, but also local geographic characteristics can be obtained by analyzing the contour trees.

Automatic generation of contour trees has been studied by a number of researchers. The principal idea of their achievements is as follows: form closed polygons for all contours, and then calculate the containment relations and parallel relations among the contours. Using these relations, the parent-child relations and sibling relations among contours can be obtained. Here presents an approach to constructing contour trees, integrating the method by Zhang et al. (2005).

3.2.1 Analysis of the Problem

Spatial relations among contours lines mainly refer to topological relations. A closed contour line can be viewed as a polygon; thus, the spatial topological relations among them include containment relation and separation relation. It is easy for people to recognize these spatial relations by their experiences, e.g. people know two adjacent contours with same elevations are separated, while two adjacent contours with different elevations have containment relations. Here, contour trees are designed to record such relations.

Now that closed contour lines are regarded as polygonal, the relations among contour lines can be viewed as the relations among polygons. Although calculation of the relations among polygons is easy in GIS (geographic information systems), construction of closed polygons by contour lines still need to be discussed.

Every contour forms a continuous curve in the real geographic world; however, some contours stop at the map borders and cannot form closed polygons. In this case the map borders need to be taken into consideration along with the contour lines to generate polygons. Nevertheless, contour lines are usually saved in a spatial database in vector data format. As far as a contour line is concerned, the order of the points on the contour line in clockwise or anti-clockwise is generally arbitrary. When the contour line is used to construct a closed polygon along with the map border, its direction (i.e. clockwise or anti-clockwise) in the map database should be known; otherwise, two different resulting polygons may be formed (Fig. 3.4). Actually, it is well known that only one contour exists in the real geographic world. Hence, correct directions of the contour lines need to be determined before the polygons are formed.

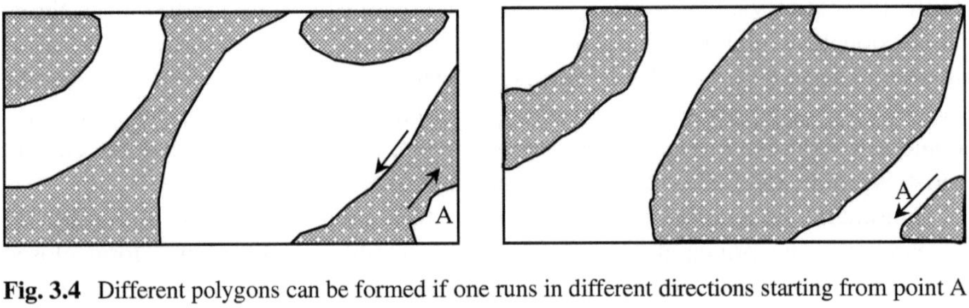

Fig. 3.4 Different polygons can be formed if one runs in different directions starting from point A

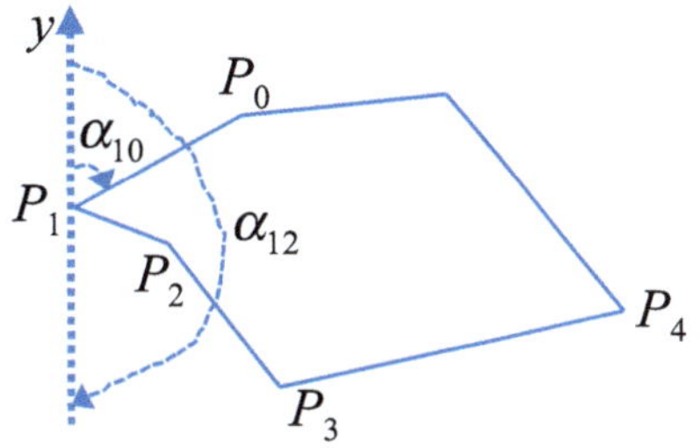

Fig. 3.5 Method for determining if the nodes of the contour line are saved in clockwise

3.2.2 Automatic Generation of Closed Contour Lines

3.2.2.1 If the Contour Line Is Closed on the Map

If a contour line is closed on the map it is simple to construct the polygon using the contour line. Suppose that the elevation of the inner of the contour line is greater than that of the outer of the contour line, it is necessary to determine if the nodes of the contour line are saved in counter-clockwise before constructing the contour trees. If it is not in counter-clockwise, the order of the nodes should be re-arranged.

A method for determining if the nodes (i.e. points) of a contour line are saved in counter-clockwise is as follows, taking Fig. 3.5 as an example: to get a point (say p_1) whose say x value is the minimum compared with the other x values of the points, connect this point with its two adjacent points, one before it (say p_0) and the other after it (say p_2). Calculate the angle (say α_{10}) formed by rotating the positive axis y to p_{10} and the angle (say α_{12}) formed by rotating the positive axis y to p_{12}. If $\alpha_{10} \leq \alpha_{12}$, it can be concluded that the nodes chain of the contour line is in counter-clockwise direction; otherwise, it is in clockwise direction.

3.2.2.2 If the Contour Line Is Not Closed on the Map

If a contour line is not closed on the map but it intersects with the map borders, supposing that the elevations at the left of the contour line is greater than that of the right and a contour line used for representing a peak is closed in counter-clockwise direction, determine and number the starting point and the ending point of each contour line according to their x-coordinates. As far as the two terminal nodes of a contour line is concerned, the terminal node with less x is called the starting point, while the other one is called the ending points. As far as a contour line is concerned, its number for the starting point is greater than that of the ending point. An example is shown in Fig. 3.6a.

Arbitrarily select and display three contour lines whose elevations are $H - \Delta H$, H, and $H+\Delta H$, respectively, where, ΔH is the contour interval of the map. Suppose that the left side of the running direction of the contours is higher than that of the right side, and all peaks are closed in counter-clockwise, here, taking Fig. 3.6 as an example, a procedure is shown below for the purpose of constructing a closed

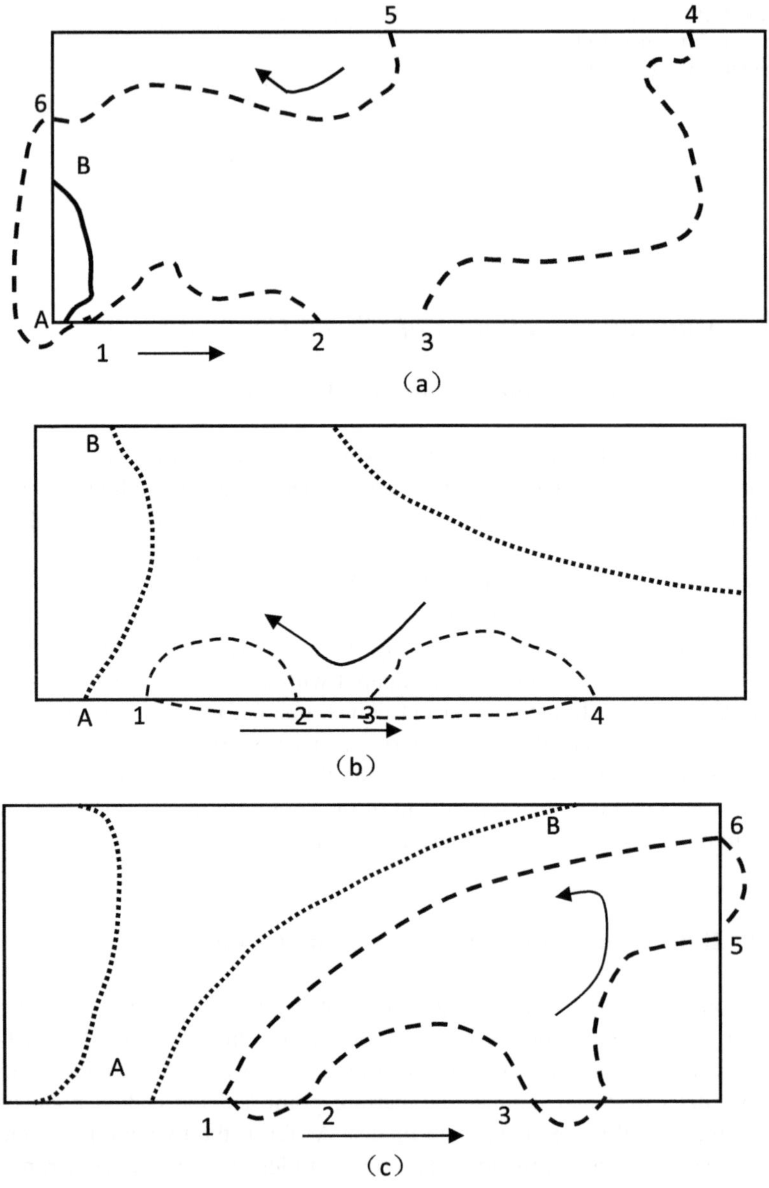

Fig. 3.6 Approach to forming polygons for the contours intersecting with the map borders

polygon for the contour lines whose elevations are *H* (they are called the current contours, represented using dash lines in Fig. 3.6a) and adjust the nodes of the current contour lines so that they are arranged in counter-clockwise direction.

Step 1: sort and number all starting and ending points of the current contour lines in the light of their x-coordinate values, e.g. they are numbered 1, 2, 3, 4, 5, and 6 in Fig. 3.6a.

Step 2: Judge if the first starting point 1 of the current contours has been tracked. If it has not, record this contour line whose elevation is H as the starting contour line.

Step 3: obtained the successive line. It is necessary to find such a contour line (i.e. the successive contour): the x-coordinate of its starting point or ending point is less than that the point 1. For example, contour AB is such as contour in Fig. 3.6a. If no successive contour can be obtained by the x-coordinates, the y-coordinates of the contour lines should be taken into consideration.

Step 4: determine the elevation relations between the current contour and the successive contour and form the polygon for the current contour (Fig. 3.6b, c).

3.2.3 Construction of Contour Trees

Contour tree is an efficient tool for describing topological relations among adjacent contour lines. Most importantly, it is easy to understand and is usually used in automated generalization of contour maps. To retrieve vector contour lines easily, contour trees should be saved using a simple linked list. Here, a group of contour lines for describing a hill or a peak can be expressed using a linked list, taking each closed contour line as a node of the linked list.

Each node can be defined using the following structure, the parent-child relations among contours can be determined by means of the containment relations among their corresponding MBRs (minimum bounding rectangles).

```
Class Ctreeleaf
  {
  public :
    int m_h;
    CLine m_leaf; //name of this node (contour line)
    CTreeleaf *m_Pnextleaf; //the next node
  };
  typedef vector <CTreeleaf> CMytree;
```

Although topological relations among contour lines can also be expressed using many other methods, contour tree has its advantages compared with those methods. For example, not only does it provide the directions of the nodes of each contour line (i.e. the nodes of the contour lines are saved in clockwise or counterclockwise), but also it tells the containment relations among contour lines; in addition, it views each closed contour line as a polygon, and forms contour trees for hills, peaks, valleys and depressions, and therefore presents fundamental topographic descriptions.

At least usage of contour trees exists at the following three aspects.

Firstly, it can be used to determine the curvatures of contour lines. If a contour line is divided into many monotonic lines, i.e. each line is either a concave or a convex curve. As is well known in the community of cartography, a convex curve denotes a ridge and a concave curve denotes a valley. To judge if a curve is concave or convex, the derivative of the middle node in three successive nodes of the curve should be calculated. Here, the contour trees can help to get correct direction of the nodes by which if the curve is convex or concave can be obtained.

Secondly, it helps to get topographic terrain lines correctly. For example, to get the valley bottom line (a curve formed by connecting the valley bottom points of the adjacent contour lines) of a group of contour lines, the valley bottom points can be obtained starting from the contour line with greatest elevation to the adjacent one with the second greatest elevation, and so on. If contour trees are employed in this process, the procedure is not only more efficient, but also more reliable in getting correct results.

Thirdly, it makes it easy to visually represent geomorphological features of the region by colored layers. Here, the most important problem is to get the elevation belts using the contour lines. Because the contour trees that describe topological relations among the contour lines can be obtained, i.e. the containment and adjacency relations among the contour lines are obtained, the elevation belts can be divided easily and correctly.

3.2.4 Detecting Topographic Terrain Lines by Means of Contour Trees

Topographic terrain lines are the configurational lines used for representing topographic fluctuation changes including ridge lines and valley lines. They can be obtained taking either DEM or contour maps as source data. The following presents an approach for detecting topographic terrain lines by means of contour trees, taking contour lines as source data (Liu and Zhan 2005). It consists of three steps: recognition of spatial relations among contour lines, calculation of topographic points, and organization of topographic terrain lines.

3.2.4.1 Recognition of Spatial Relations Among Contour Lines by Contour Trees

There are four types of spatial relations among contour lines, i.e. direction relation, distance relation, topological relation, and similarity relation. However, direction, distance and similarity relations among contour lines are usually not that useful. Therefore, only topological relations among contour lines are considered here while spatial relations are discussed.

Contour lines have the following characteristics: (1) all points on the same contour lines own the same elevation; (2) every contour line is a continuous and closed curve; (3) all contour lines are measurable; (4) contour lines exist in groups, and they are similar in graphics and do not intersect with each other. It is easy to find that contour tree is a good tool for describing the spatial topological relations among contour line according to their above characteristics (Boyell and Ruston 1963).

A contour tree is a directed tree. Every node of the tree corresponds to a contour line. Its root is the contour line with the least elevation value. Each node has an only parent node, and a parent node may contain many children nodes and forms a branch of the tree. This tree not only can denote the elevation values of all contours, but also can tell the relations of their elevation values and spatial topological relations including containment relations and adjacency or parallel relations.

After the contour tree is constructed, the spatial relations among the contour lines can be detected abiding by the following four principles.

(1) Suppose that the root is at the level 0, the level difference between a parent node and one of its direct children nodes is 1. An edge connecting two adjacent nodes is called a path. The number of the edges between two nodes is called the connection degree of the two nodes. It is obvious that the connection degree between a parent node and one of its direct children nodes is 1.
(2) If the connection degree between two nodes is equal to their level difference value, the two contour lines corresponding to the two nodes have topologically n-order inclusive. Here, n is the level difference between the two nodes.
(3) If two nodes in the tree are at the same level and the connection degree between each of two nodes and their parent node is 1, the connection degree of the two nodes is 2. So, it can be concluded that if any two nodes at the same level and their connection degree is 2, the two contour lines must be topologically parallel (in other words, they are adjacent brothers).
(4) If the connection degree between two nodes is not equal to the level difference between them, the two corresponding contour lines are topologically n-order disjoint. Here, $n = k - 2$ and k is the connection degree of the two nodes.

3.2.4.2 Calculation of Topographic Feature Points

Topographic feature points are the key points on contour lines which form the "skeletons" of contour lines. They are usually ridge points and valley points, and they can be obtained by the following two steps.

Preprocessing Contour Lines by Convex Hulls

It is easy to get a group of contour lines with inclusive relations, because these contour lines have parent-child relations in the contour tree. Generally, the contour lines with inclusive relations are similar in shape. If the convex hull of each of the contour lines is constructed, all the convex hulls are also similar in shape (Fig. 3.7a).

Given a contour line $L = (p_0, p_1, \ldots\ldots, p_n)$, its convex hull is $C = \left(p'_0, p'_1, \ldots, p'_m\right)$, where is the number of nodes on the contour line, and is the

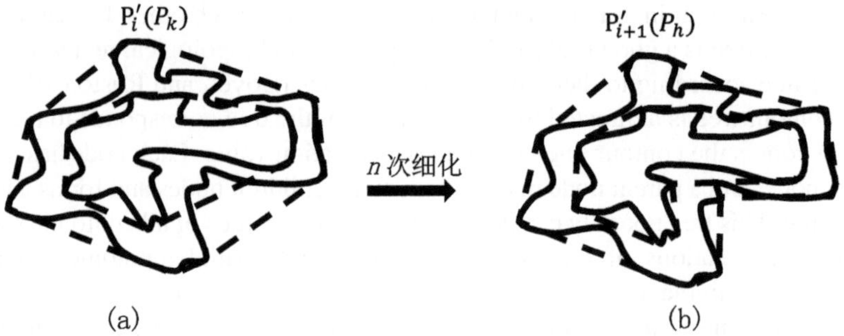

Fig. 3.7 Constructing convex hulls for a group of contour lines with inclusive relations in the contour tree

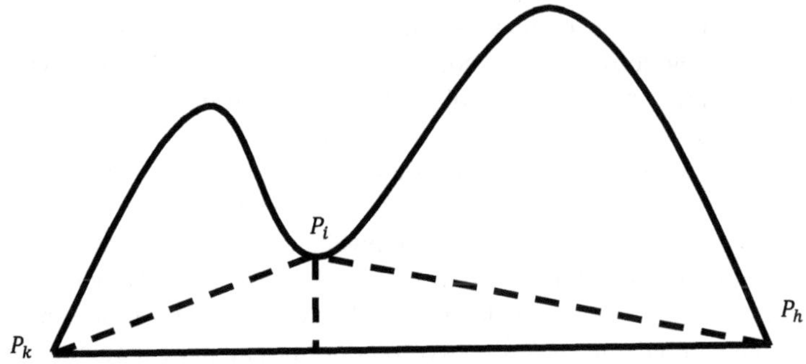

Fig. 3.8 Principle for thinning convex hulls

number of vertices on the convex hull. Let $p_i' = p_k$ and $p_{i+1}' = p_h$. Here, $p_i' \in \mathrm{L}$ and $p_{i+1}' \in \mathrm{L}$.

If $|h - k| > 1$, line segment $p_i'p_{i+1}'$ is called an edge of the convex hull. The curve formed by connecting the points p_k between and p_h on the contour line is called a curvature of the convex hull.

As far as a group of contour lines are concerned, there are a group of corresponding convex hulls. Given a distance tolerance value D_T, compare each edge of the convex hulls with D_T. If any edge of the convex hulls is greater than D_T, the group of the convex hull need to be thinned until every edge of the convex hulls is less than D_T. This procedure is the same as that of the Douglas-Peucker Algorithm (Ramer 1972; Douglas and Peucker 1973), but its constraint is different from that of the Douglas-Peucker Algorithm.

Taking Fig. 3.8 as an example, $(p_k, \ldots, p_i, \ldots, p_h)$ is a curvature of a convex hull, and p_kp_h is an edge of the convex hull. Point p_i is the point on the curvature who has the nearest distance to p_kp_h. Connect p_i and p_k, and p_i and p_h, and two new edges of the convex hull can be got, i.e. p_kp_i and p_ip_h. If each of the new edges is less than the tolerance value, stop the thinning procedure. Otherwise, the above procedure should be recursively executed. A resulting example of the thinning procedure is shown in

Fig. 3.7b. Actually, the procedure for thinning convex hulls is for dividing contour lines. It expresses the inclusive relations among contour line using the inclusive relations among the curvatures of the convex hulls, which facilitates quantitative calculation of their spatial relations.

Obtaining Topographic Terrain Points by the One-Step Dougas-Peucker Algorithm

Connect the first and the ending nodes of each curvature of the contour lines. Each of the curvature is formed by the edges of the convex hulls. Detect each curvature by the so-called one-step Douglas-Peucker Algorithm, which means the Douglas-Peucker Algorithm only search one time for the farthest point (say, P_f) from the curvature to the line segment connecting the first and the ending nodes of the curvature. It is easy to draw a conclusion that P_f is a topographic terrain point on the contour line. If the curvature is corresponding to a ridge, the point is a ridge point; otherwise, if the curvature is corresponding to a valley, it is a valley point. By this way, all topographic terrain points can be obtained.

3.2.4.3 Organization of Topographic Terrain Lines

Although the ridge points and valley points can be obtained, it is unknown which points belong to the same ridge or valley. Here, 'similar triangles' are used to determine how to connect topographic terrain points to form topographic terrain lines.

Convex hull triangle is defined: it is a triangle (say, T) formed by connecting the two ending points of an edge (say, E) of a convex hull with the corresponding topographic terrain point (say, P). P is called the vertex of T, and E is called the based edge of T.

Three factors need to be considered in evaluating if two convex hull triangles (T_1 and T_2) are similar triangles.

- Distance between the two triangles (say, d): it refers to the distance between the two vertices of the two convex hull triangles.
- Orientation or direction between the two triangles (say, o): supposing that the line connecting the vertices of T_1 and T_2 is L, the angle between L and the bisector of the vertical angle of T_2.
- Parallel factor (say, p): it refers to the angle between the base edge of T_1 and the base edge of T_2.
- If the following requirements can be satisfied, T_1 and T_2 are viewed as similar triangles.
- d s less than the tolerance value (it is a given value).
- L intersects with the base edge of T_2, and should be less than the tolerance value (it is a given value).

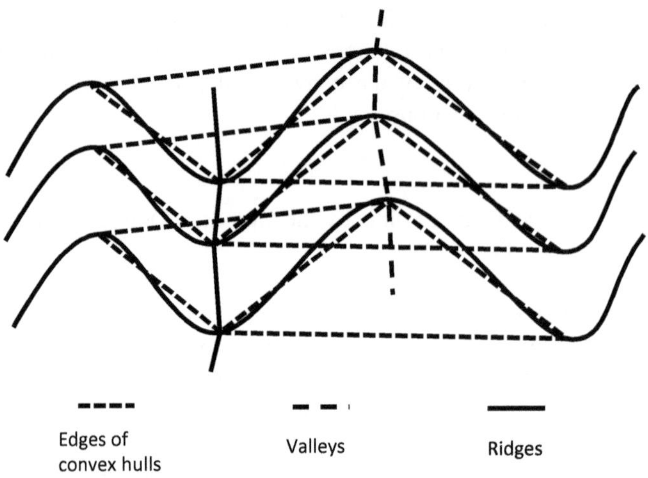

Fig. 3.9 Approach to detecting topographic terrain lines

- The two base edges of the two convex hull triangles are approximately parallel, which means is p less than a given tolerance value.

If T_1 and T_2 are similar triangles, their two vertices belong to the same topographic terrain line. In other words, the two vertices can be connected to form a topographic terrain line.

It is obvious that ridge lines and valley lines are dual graphs on the plan. If there must exist a ridge line between two adjacent valley lines and vice versa. Werner (1988) named this 'interlocking structure'. Interlocking structures formed by ridge lines and valley lines are useful in recognizing topographic terrain lines, because if one type of the lines is obtained, the other type of the lines can be easily obtained by similar methods.

Figure 3.9 shows the approach for capturing and organizing topographic terrain lines.

3.3 Description of Contour Lines by Topographic Terrain Lines

Peaks, ridges, valley and depressions are key landforms that contain the frameworks of earth surface, and play important roles in making maps, relief recognition, hydrological analysis, construction of digital elevation models (Mark 1984; Yeoli 1984; Band 1986; Weibel 1992). Achievements in quantitative analysis of ridges and valleys can be dated back to Horton (1945). After that applications are extended to geo-statistics and geographically topological analysis (Strahler 1952; Scheidegger 1996; Shreve 1996). Currently, three types of data are used to record topography, i.e. contour lines, digital elevation models in GRID format, and digital elevation models in TIN format. Topographic information is hidden in the three types of data and need to be extracted when necessary; thus, many approaches have been proposed to detect topographic information (O'Callaghan and Mark 1984; Werner 1988; Wolf 1988; Qian et al. 1990; Tang 1992). The following presents an algorithm for extracting and expressing topographic information taking contour lines as original data (Ai 2007). The basic idea of the algorithm is: triangulate the vector contour lines, and then extract topographic terrain lines from the triangular network.

3.3.1 Analysis of the Curvatures of a Singular Contour Line and Construction of Its Binary Tree

Contour line is a technique that represents the 3-dimensional geographic land surface on a 2-dimensional plan. Typical landforms, such as ridges and valleys, are expressed using groups of geometric curves. As far as a singular contour line is concerned, a subsection of the contour line can be viewed as the projection of a part of the corresponding valley onto a plan. Therefore, if every subsection of the valley is obtained, the whole structure of the valley can be got by connecting these subsection s of the valley in order with the help of water flow directions and grade directions of the slope. So, it is obvious that the key of this issue is "how to divide the contour line to get a number of subsections of the valley" (Mark 1984; Jones et al. 1990).

Firstly, the run of the contour line is adjusted to make sure that its right is higher than its right in elevation. This can be obtained by changing the order of the coordinate chain of the contour lines.

Secondly, construct Delaunay triangular network of the contour line, taking the vertices of the contour line as the point set.

Thirdly, select the triangle at the same side of the valley, and construct a binary tree (Ai et al. 2000) to express the containment relations among the curvatures of the contour line (Fig. 3.10).

3.3.2 Selection of Subsections of Valleys

In the light of fractal geometry, a valley can be infinitely divided into small curvatures. Thus, extraction of topographic terrain points should consider the resolution of the curvatures. Generally speaking, each node of the binary tree is corresponding to the projection of a subsection of the contour line at a specific resolution. For example, in Fig. 3.10, node G corresponds to a subsection of the valley at a specific resolution. If the topography at higher resolution is needed, node

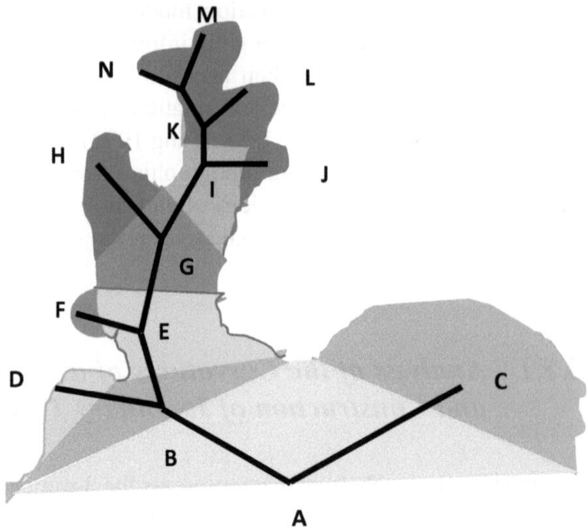

Fig. 3.10 Expressing topological relations among the curvatures of the contour line by binary tree

G can be substituted by nodes H and I. Furtherly, node I can be substituted by J and K if more detailed topography is needed.

If the binary tree of the curvatures of a valley is formed, subsection s of the valley at different resolutions can be obtained by searching the binary tree taking given depth of the tree as the parameter for controlling the searching procedure. Supposing that the tolerance value of the depth of the valley is λ, search the binary tree starting from the root who owns the greatest depth of the tree. If coming across a node whose depth is less than λ, stop the procedure and select the parent node of this node. The corresponding subsection of the valley is the required subsection of the valley. A special condition that needs to be considered is: when reaching a leaf node and its depth is still less than λ, this node should be selected to keep the wholeness of the valley.

Figure 3.11 shows three different subsets of subsections of the same valley obtained by searching the binary tree using different depth values.

3.3.3 Detection of Sharp Peaks of Valleys

Some researchers detected sharp peaks of valleys by curvatures of curves (Wolf 1988), and defined the sharp peak of a valley as the point with the greatest curvature at one side on the contour line. However, because vector data is most sensitive to coordinate changes, such as "shaking" in digitization, the results obtained by calculating curvatures generally are not satisfactory. Detection of sharp peaks of valleys needs to consider not only local curvature changes but also the context of the contour line.

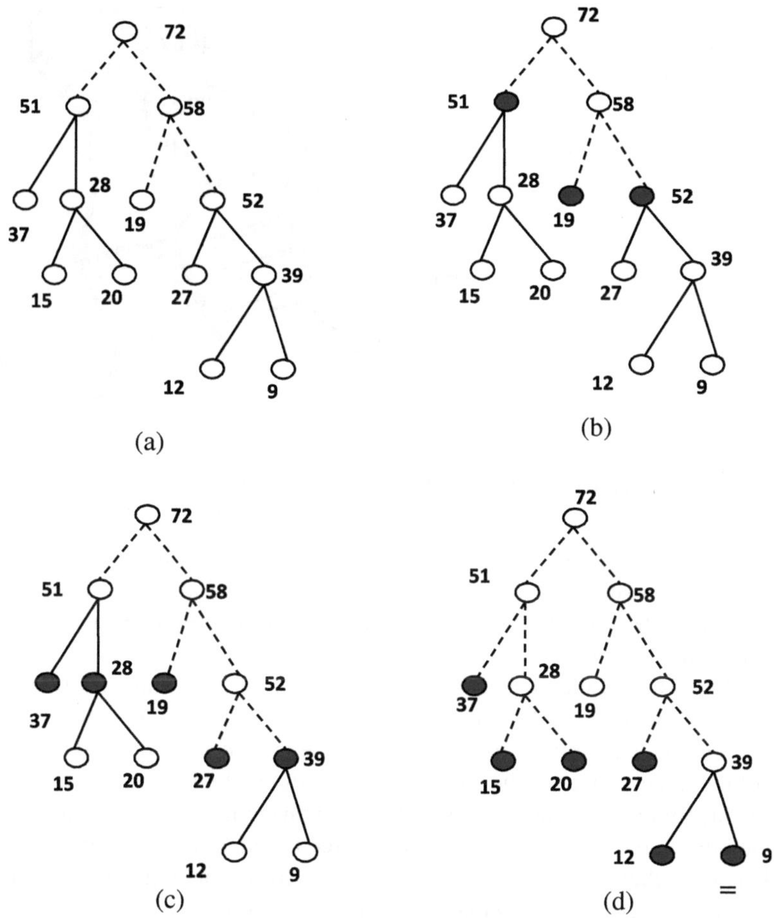

Fig. 3.11 Searching results of binary tree for describing the subsections of the valley. Dashed line denotes a connection is sheared off; the digit beside each node is the depth of the node; the λ is the tolerance of depth; black dotted nodes are the selected ones. (**a**) Original binary tree. (**b**) Selected nodes when $\lambda = 50$. (**c**) Selected nodes when $\lambda = 35$. (**d**) Selected nodes when $\lambda = 20$

To fill the above gap, Ai (2007) proposed a Delaunay triangulation-based approach for detecting sharp peaks of valleys. This approach considers both local curvatures and its context in the contour line, and the results obtained by the approach are visual acceptable. The basic idea of the approach is: when selecting a subsection of a valley, detect the triangles contained by the subsection, and extract skeletons and form a skeleton tree whose root is at the opening of the valley. Compare the distance from each node to the root and select the node with the longest distance. This node is the sharp peak of the valley.

For example, in Fig. 3.12, compare the skeleton distances from root O to the four nodes A, B, C and D, and obtain that C is the sharp peak of the valley.

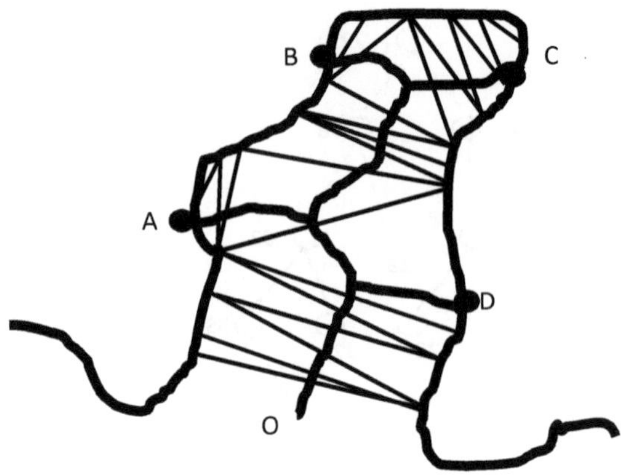

Fig. 3.12 Determination of the sharp peak by the longest branch of the skeleton tree

3.3.4 Construction of Plan Structural Trees

Because each valley consists of many also sub-valleys and each sub-valley contains many smaller sub-sub-valleys, each valley may be viewed as a tree. Thus, a valley tree can clearly describe the water flow direction and the relations among its main stream and the branches.

Hydrologically a valley has its main stream and branches, and each branch may contain many smaller branches that flow into it. If the branches are defined as nodes of a tree, and the containment relations among the branches are defined as "connections" of the nodes, a tree of the valley in semantics can be formed and it is called a semantic hierarchical tree of the valley. To distinguish the semantic hierarchical tree from the binary tree of the valley, the binary tree is called the plan structural tree.

Taking the valley in Fig. 3.13 as an example, its corresponding plan structural tree is shown in Fig. 3.14, and its semantic hierarchical tree is presented in Fig. 3.15.

A plan structural tree presents the topological relations of the branches of a valley in a plan, while a semantic hierarchical tree of the valley expresses the differences of the branches in geography and semantics. The latter is a detailed expression of the former, and it answers: "which one is the main or major stream of the valley". Correspondingly, to some extent, Horton (1945) also answered this question in his coding system.

A plan structural tree is the connection of the vertices of the valley in the light of their adjacency relations. The following three steps are needed to form a plan structural tree of the valley.

Step 1: connection of the sharp peak

To determine how to connect the sharp peaks, a couple of important constraints should be considered, such as the distance between two adjacent sharp peaks (Fig. 3.16a), directions of the norms of the line connecting two adjacent sharp peaks (Fig. 3.16b), and if the line connecting the two adjacent sharp peaks intersects

Fig. 3.13 A valley system

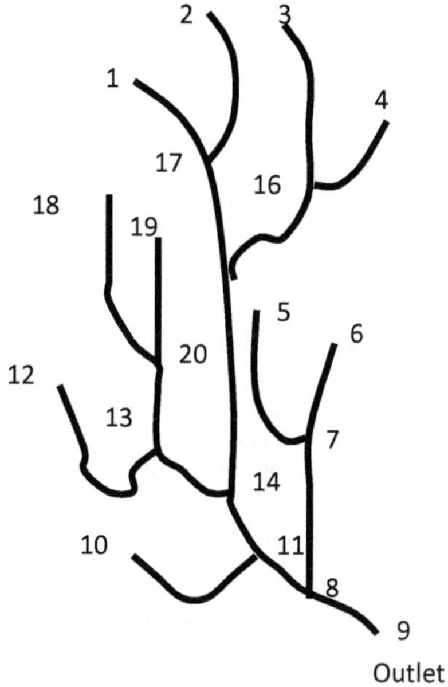

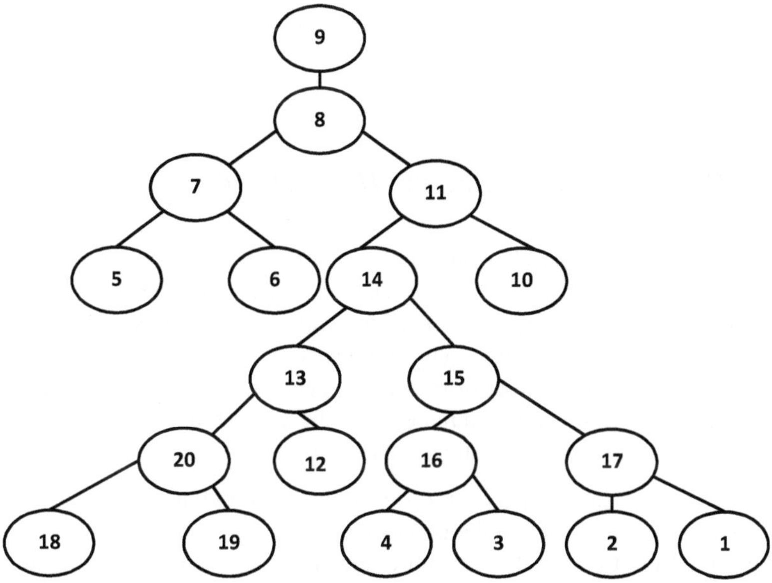

Fig. 3.14 A plan structural tree for describing topological relations among the branches a valley

with the open of the open of the valley (Fig. 3.16c). How to assign appropriate weights to these constraints is a key step which depends on human's experiences.

Step 2: connection of the branches

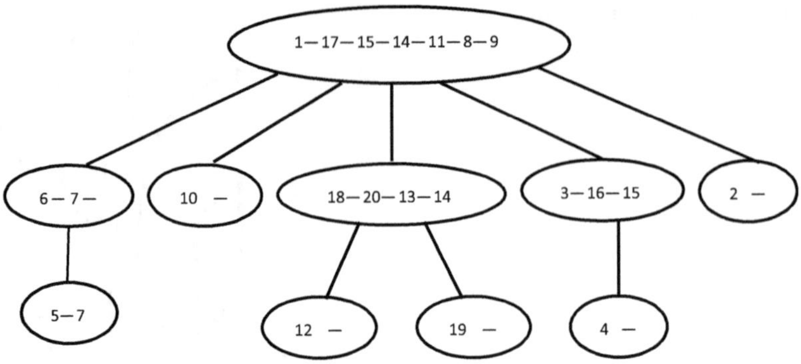

Fig. 3.15 A semantic hierarchical tree for describing the connection relations among the main stream and the branches of a valley

To form a tree of the valley, the small branches connected by the sharp peaks need to be furtherly connected to establish their topological relations (mainly neighboring relations among the branches).

Step 3: re-organization of the valley trees

Check the topological relations of the branches, such as if they are connected correctly; then construct the plan structural tree of the valley by matching the nodes with branches and water flow directions.

3.3.5 Organization of Semantic Hierarchical Trees

Select an appropriate branch from the plan structural tree and generate a main stream, and then set up the confluence relations among the branches and the main stream. This is really a re-organization of the valley in semantics.

According to water flow directions, confluence of a couple of branches from the upstream at a vertex of the plan structural tree generates an only main stream of the downstream. If the following requirements can be met, a branch of the upstream can be viewed as a part of the main stream that connect with an only downstream:

(1) it is continuous in geometry with some downstream branches;
(2) it owns longest distance to the downstream;
(3) the slope from the branch to the downstream is not steep compared with that of its adjacent branches;
(4) its Horton code value is greater than its neighboring branches.

Fig. 3.16 Geometric constraints in connecting valleys (**a**) the shortest one (AC < AB)is not the valley line (**b**) the straight open is not the valley line (**c**) the line intersecting with the open is not the curve line

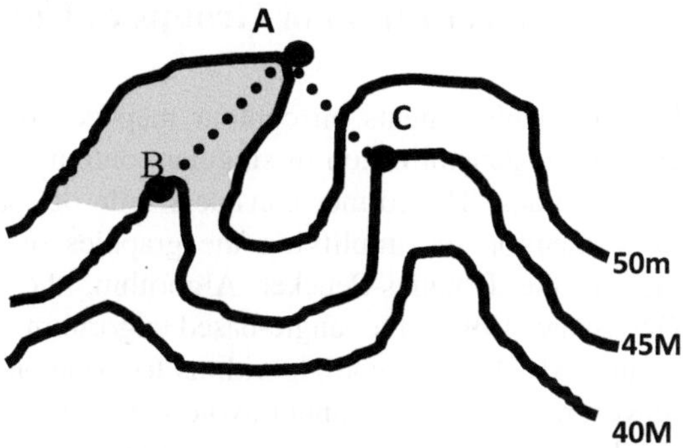

(a)

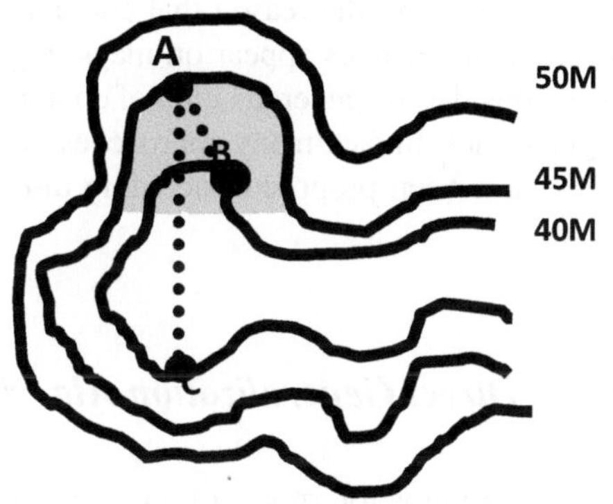

(b)

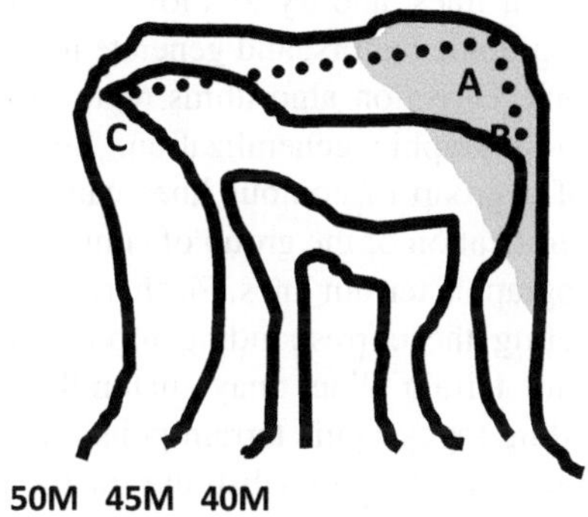

(c)

3.4 Generalization of Groups of Contour Lines

Existing achievements on contour map generalization can be classified into two categories: generalization of singular contour lines and generalization of groups of contour lines. The former considers little of spatial relations among contours but pays attention to simplifying the graphics of contour lines. Typical algorithms include the Douglas-Peucker Algorithm, the Li-Openshaw Algorithm (Li and Openshaw 1992), the angle-based algorithm and the distance-based algorithm (Yan et al. 2007). Some algorithms try to avoid self-intersection of contour lines; however, they usually cannot avoid the simplified contour line intersecting with its neighboring ones, because these algorithms do not take into account spatial relations among contour lines in the process of generalizing contour lines.

As is well known, the reason that contour lines can express topographic terrain well is that contour lines appear on maps in groups (Richardson 1961). Thus, it is natural to consider the generalization of contour lines by viewing them as groups but not singular lines. In fact, many approaches to simplifying contour lines taking them as groups have been proposed, including direct algorithms and indirect algorithms (Fei 1998).

3.4.1 A Direct Generalization Algorithm

Direction algorithms for groups of contour line generalization extract the structures, configurations, spatial relations, and topographic terrain lines of the will-be-generalized contour lines, and try to know the relations among the contour lines and the above topographic factors, and generate new contour lines on smaller scale maps by the relations. Direction algorithms regard topographic terrain lines as an essential factor in topographic generalization. Because the curvatures, configuration and patterns of a group of contour lines can be describe using its topographic terrain lines, generalization of the group of contour lines may be equal to the generalization of the topographic terrain lines. Further, because a topographic terrain line is formed by connecting the corresponding topographic terrain points, generalization of a topographic terrain line may naturally be equal to the generalization of corresponding topographic terrain points. In conclusion, generalization of the topographic terrain points can substitute for the generalization of the group of contour lines.

Here presents a typical direction algorithm. Its main idea is as follows: firstly, the contour line data are preprocessed; secondly, ridge points, valley points and other key points on the contour lines are recognized and recorded; after this, topographic terrain lines are tracked and formed; lastly, the contour lines are generalized. To simplify the discussion of the algorithm, it is supposed that all studied topographic landforms are positive.

(1) Preprocessing of the contour line data

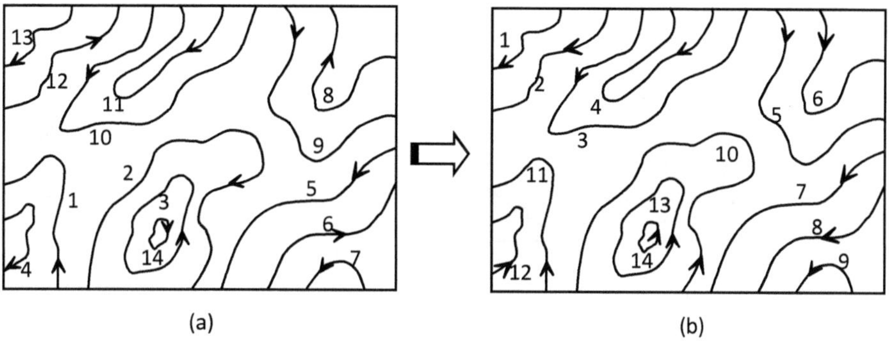

Fig. 3.17 Preprocessing of the contour line data

Supposing that all contour lines of a map are saved in file, every contour line forms a continuous chain whether it is a closed line or an open line; but it is uncertain for each of the contour lines that if the corresponding chain is save in clockwise or counter-clockwise order (Fig. 3.17a).

Search for the ending points of all the contour lines clockwise starting from the northwest corner of the map. When an ending point of a contour line is got, the contour line is recorded, and the run of the contour line is adjusted as counter-clockwise by changing the order of the nodes saved in the file if the original run of the contour line in the file is clockwise.

After a round of the search, all open contour lines are adjusted. After this, the remaining contour lines, i.e. closed contour lines, are adjusted, too. If the run of a closed contour line is clockwise, change the order of their nodes to make the contour line become counter-clockwise; otherwise, keep it as it is.

(2) Detection of the key points

If the curvature of a point of a contour line is greater than a given tolerance value, this point can be regarded as a key point of the contour line. A formula of a continuous contour line can be expressed as follows:

$$\begin{cases} X = P_0 + P_1 Z + P_2 Z^2 + P_3 Z^3 \\ Y = Q_0 + Q_1 Z + Q_2 Z^2 + Q_3 Z^3 \end{cases} \quad (3.1)$$

By this formula, any specific continuous contour line has a specific similar formula with different parameters. Therefore, it is easy to get the curvature of each points on the contour line, and further the key points of the contour line can be obtained.

(3) Extraction of topographic terrain points

As far as a specific contour line is concerned, not all the topographic terrain points are included in the key points. Therefore, after the key points are obtained, the topographic terrain points (i.e. ridge points and valley points) also need to be

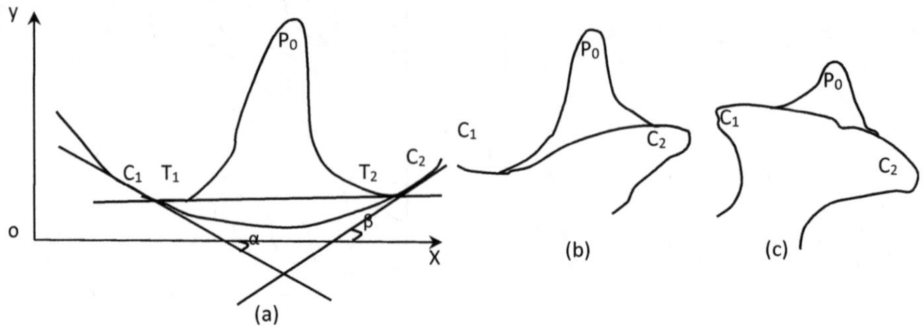

Fig. 3.18 The greatest angle-based iterative algorithm for generalizing contour lines

extracted from the contour line. After they are obtained, they are contrasted with the key points and the two sets are combined to form a new point set (say, S) which contains all the points needed.

(4) Track of topographic terrain lines

Starting from a contour line with the least elevation value, find a valley/ridge point (say, p_0) starting from one of its two ending points; then search for its most adjacent contour line with greater elevation value, and detect the valley/ridge point (say, p_1) at the same valley/ridge with p_0. Iteratively execute this procedure until no valley/ridge point at the same valley/ridge can be found. Sequentially connect the valley/ridge points starting from p_0 and a valley is constructed.

Repetitively execute this procedure until no new valley point and ridge point can be found.

(5) Generalization of contour lines

The contour lines may be generalized by calculating the greatest angle. Taking Fig. 3.18 as an example, firstly the common tangent points T_1 and T_2 on the contour line are obtained; then a point C_1 adjacent to T_1 and a point C_2 adjacent to T_2 are selected (Fig. 3.11a); lastly, a new and simpler curve between C_1 and C_2 are generated by the formula 3.1, and it substitutes for the old curve segment between C_1 and C_2.

Track all the contour lines by this method, and they can all be simplified.

3.4.2 An Indirect Generalization Algorithm

Indirect generalization algorithms do not directly simplify contour lines. On the contrary, they simplify the DEMs (digital elevation model) corresponding to the groups of contour lines. The idea of these algorithms is: contour line is a tool for describing topographic fluctuations, and so is DEM. Additionally, DEMs and contour lines can generate each. So, generalization of contour lines of a region is corresponding to generalization of DEMs of the region. If simplified DEMs of the region are obtained, they can be easily transformed to get the contour lines of the region.

The following presents an indirect generalization algorithm proposed by Zhu et al. (2008). This algorithm firstly changes contour lines to DEMs; then it generalizes the DEMs and makes it become another DEMs at specific smaller scale; lastly, it generates the contour lines at the specific smaller scale by the new DEMs. Supposing that a topographic map containing many groups of contour lines are given, the algorithm can be described by the following three steps.

(1) Construction of the TIN

Many approaches for generating TIN (Triangular Irregular Network) using contour lines have been proposed in literature, and almost any one of them may be employed in this algorithm. That needs to be pay more attention to is the data structure used for recording the vertices, edges, and their relations. A good data structure not only can make the algorithm simple to program, but also greatly improve the efficiency of the algorithm. Here presents a commonly used data structure for recording triangles of TINs.

```
Class CTriangle
  {
  public:
     int ID;                  //ID of this triangle
     int ID-vertex1, ID-vertex2, ID-vertex3; //three IDs of the vertices
  of the triangle
     int ID-edge1, ID-edge2, ID-edge3;    //three IDs of the edges of the
  triangle
    CTriangle *neighbour1, *neighbour2, *neighbour3; //three neighbours
  of the triangle
  };
```

(2) Simplification of the TIN

TIN simplification has been studied for years. The following addresses the Progressive Mesh-based Method (Hoppe 1996) and the Maximum Independent Point Set Method (Zhang 2002).

Progressive meshes (Hoppe 1996) are continuous multi-resolution representations of arbitrary meshes. The meshes can be simplified by iteratively folding their edges. For example, Fig. 3.19a is the original meshes whose one edge will be folded; Fig. 3.19b is the meshes after the edge is folded. The technique of the edge folding is as follows: change the will-be-folded edge to a point (say, P_N); in the meanwhile, delete the triangles (one or two triangles) that contain the will-be-folded edge, and re-triangulate the areas that are directly adjacent to this edge. P_N may the midpoint or one of the two vertices of the will-be-deleted edge; it may also be other point, such the centroid of the quadrangle containing the will-be-deleted edge.

(a) (b)

Fig. 3.19 The progressive mesh-based method (**a**) the thicker edge will be folded (**b**) the graphics after edge folding

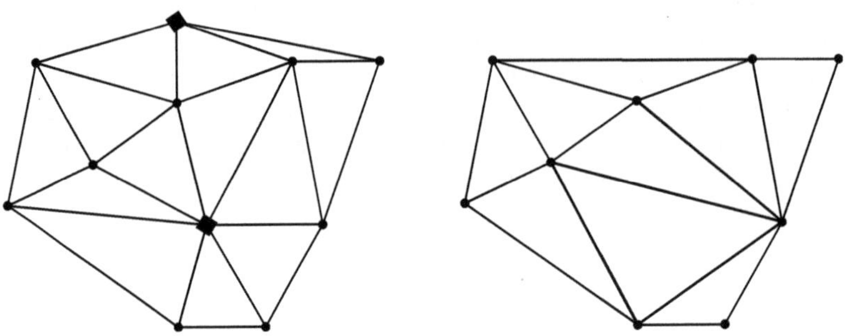

(a) (b)

Fig. 3.20 Reconstructing the hierarchical topography by the Maximum Independent Point Set Method (**a**) the two will-be-deleted points (the quadrangle points) (**b**) re-triangulation

The idea of the Maximum Independent Point Set Method (Zhang 2002) for constructing hierarchical topography is as follows: starting from the most detailed level of the network, delete the most unimportant vertex (i.e. the most independent vertex) and get a new triangular network of the studied area. It should be noticed that if the deleted vertex is a topographic terrain point (i.e. a ridge point or a valley point) it is not favorable to maintain the topography; therefore, in practice the topographic terrain points are not allowed to be deleted.

Here, gives the definition of independent point set: if two vertices of a TIN are not neighbours, they are called independent points. Further, if a set owns the maximum vertices of the TIN and no any two vertices of the set are neighbours, this set is called the maximum independent point set of the TIN.

Supposing that DT_i is the most detailed topography of the studied area, its corresponding triangle is D_i. The point set including all vertices of all the triangles of the TIN is V_0. The procedure of the Maximum Independent Point Set Method can be described as follows in C++ (Fig. 3.20):

Step 1: let i = 0.

Step 2, starting from DT_i, search for the maximum independent vertices (say, I_i), and remove I_i from V_i, and then get the point V_{i+1} set that represents the topography of the studied area at coarser level.

Step 3, triangulate of the studied area taking V_{i+1} as the point set. Let i = i+1. Search for the most detailed topography in the new TIN and recorded it as DT_i.

Step 4, if the new TIN can meet the requirement, stop the procedure; else, go to step 2 and go on with the procedure.

(3) Generation of contour lines

If a specific elevation value is given and the TIN of a given area is searched and interpolated to form contour lines, usually three cases need to be considered when a contour line is being tracked:

- if the specific elevation value is not within the elevations of the three vertices, the contour line does not intersect with this triangle.
- if the specific elevation value is equal to the elevation of one of the three vertices, the contour line pass through that vertex.
- otherwise, the contour line intersects with the triangle. Here, we have two conditions to deal with. If the specific elevation value is equal to the elevation of one of the three vertices, the contour line passes through the vertex and intersects with the opposite edge of the vertex; otherwise, the contour line passes through an edge of the triangle, and intersect with one of the other two edges whose elevation difference is greater.

3.4.3 Discussion of the Algorithms for Contour Map Generalization

Generalization of singular contour lines (e.g. the Douglas-Peucker Algorithm (Douglas and Peucker 1973)), belonging to direct generalization techniques, simplifies curvatures of each contour line. If studied area is flat and the original map belongs to large scale it has many advantages. For example, it can quickly simplify the shape of the contour lines, greatly reduce the amount of the data while retaining the characteristics of the lines. Additionally, its execution speed is usually fast, and the algorithms are easy to be programmed. On the contrary, if studied area is a topographical complex and complicated region and/or the change of map scale is comparatively great, such an approach may not generate ideal results. For example, it might lead to self-intersections of the simplified contour line and present incorrect or false topography. Although some algorithms can avoid self-intersection of the generalized contour line, they usually do not consider the intersections among adjacent contour lines and need additional approaches to deal with.

Traditionally, the simplification procedure of singular linear cartographic objects (for example coastlines), which are represented by the same width line symbols at all scales, is supported by an empirical law called the Principles of Selection (Töpfer and Pillewizer 1966). It may be mathematically expressed as follows (Jones and Abraham 1987):

$$N_r = N_s \frac{M_r}{M_s} \tag{3.2}$$

Where, N_r and N_s are the numbers of line segments at the result and source scales, and: M_r and M_s the result and source scales, respectively.

In fact, the Principles of Selection can help the cartographers to estimate the number of the retained vertices of the simplified cartographic line (McMaster 1986, 1987). This empirical law answers to the query "how many objects should be retained on the result map", and does not provide any information about the query "which objects should be retained (McMaster 1989). In addition, the Principles of Selection are based only on the scales ratio (of the source and result map) and do not take into account the line complexity, which means that two linear objects with different degree of line complexity are treated exactly in the same way (Nakos 1990). The first disadvantage of the Principles of Selection may be overcome by applying existing line simplification algorithms, based on geometric criteria, which preserve the shapes of linear objects, like the Douglas and Peucker (1973) or Visvalingam and Wyatt (1993) algorithms. A functional description of the above-mentioned algorithms can be found in Weibel (1997). The second disadvantage may be overcome by introducing fractal geometry and more specifically the property of self-similarity (Mandelbrot 1967, 1975, 1982, 1985; Mark and Aronson 1984) into line simplification procedure. Nakos (1990) proved that for self-similar linear cartographic objects, the Principles of Selection can be expressed as follows:

$$N_r = N_{s\left(\frac{M_r}{M_s}\right)^D} \tag{3.3}$$

where: D is the fractal dimension of the linear cartographic object.

Nakos (1990) proposed a methodology for applying fractal geometry in line simplification procedure with three statistical tests, which are based on the introduced methods of estimating the fractal dimension. One can simplify the linear cartographic object by applying a line simplification algorithm by preserving the number of vertices, which have been calculated with the help of the estimated fractal dimension D.

The fractal-based method (Nakos 1990, 1996, 1999) is a kind of direct generalization approaches. It is obviously superior to the Douglas-Peucker Algorithm in processing complex contour lines, because it can consider the differences of all processed contour lines in expressing terrain and deal with the contour lines using different indices (i.e. fractal dimensions). Thus, it may be viewed as an intermediate method between singular contour line generalization methods and the methods that regard contour lines as groups.

It is groups of contour lines that can describe relief and terrain, therefore processing of contour lines in groups but not singular contour lines is an effective way to generalize topography. Topography can be generalized using direct and indirect methods.

An indirect method usually firstly transfers original contour line data into a type of digital elevation model such as GRID or TIN at a specific resolution and data format; then the digital elevation model is generalized according to designated resolution and geographic features of the studied area, and obtain a new digital elevation model; lastly, the new elevation model is changed to get resulting contour lines. The advantages of the indirect method are: the intermediate digital elevation models not only can be used to generate contour lines, but also may be used to generate some by-products such as profiles, slope figures and three-dimensional terrain graphs that are useful in various spatial analyses. On the contrary, the indirect methods have their disadvantages: information containing in the original contour lines might be lost in the process of transferring the contour lines into the digital elevation model; and then the information might be furtherly lost in the process of generalizing the digital elevation model to get the new digital elevation model at lower resolution. So, the resulting contour lines generated by the resulting digital elevation model at lower resolution are usually at low accuracy.

The direct methods generate resulting contour lines directly from original contour lines, without transferring them into other formats. Their advantages are: original information containing in the contour lines are used efficiently, and the resulting contour lines are at high accuracy. Their disadvantage is: no intermediate by-products can be generated.

3.5 Summary of the Chapter

Contour lines are a widely utilized representation of terrain models in both cartography and Geographical Information Systems. Since people often need to view fluctuations of the Earth's surface at different scales, there is a need for generalization techniques in generating multi-scale contour lines.

Mathematical description of contour lines is the foundation of generalization approaches. Therefore, after giving a short introduction of topography and contour line, this chapter spends more words on this aspect and addresses in detail the concepts, construction and applications of contour trees firstly, and then it discusses a method for Description of contour lines by topographic terrain lines; after this, two kinds of algorithms for contour map generalization are presented, i.e. indirect methods and direct methods. Lastly, it gives an overview of existing algorithms for singular and groups contour lines generalization.

References

Ai T., 2007, The drainage network extraction from contour lines for contour line generalization, ISPRS Journal of Photogrammetry & Remote Sensing 62: 93–103

Ai T., Guo RZ, Liu YL. 2000, A Binary Tree Representation of Bend Hierarchical Structure Based on Gestalt Principles. Proceedings of the 9th International Symposium on Spatial Data Handling. Beijing, China, 2000.

Band L.E., 1986, Topographic Partition of Watersheds with Digital Elevation Models, Water Resource Research, 22(1):15–24.

Boyell R.L. and Ruston H., 1963, Hybrid techniques for real-time radar simulation. In the Proceedings of the 1963 Fall Joint Computer Conference, pages 445–458.

Cheung, C. K., Shi, W. (2004): Estimation of the Positional Uncertainty in Line Simplification in GIS. The Cartographic Journal, Vol.41, No. 1, 37–45.

Cromley, R. G. (1991): Hierarchical Methods of Line Simplification. Cartography and Geographic Information Systems, Vol. 18, No.2, 125–131.

Douglas, D.H., and Peucker, T.K., 1973. Algorithms for the Reduction of the Number of Points Required to Represent a Digitized Line or Its Caricature. The Canadian Cartographer, 10,2, 112–122.

Fei L.f, 1998, Experiments on the group generalization of contour lines for topographic maps, Geo-spatial Information Science, 1:1, 85–95.

Hoppe H. 1996, Progressive Meshes. In: ACM SIGGRAPH'96 Proceedings, New York, USA

Horton R E. 1945, Erosion Development of Streams and Drainage Basins: Hydrophysical Approach to Quantitative Morphology. Bulletin of the Geological Society of America, 56(3):275–370.

Joao, E. M. (1998): Causes and consequences of map generalisation. Research Monographs in Geographical Information Systems. Taylor & Francis, London.

Jones N L, Wright S G, Maidment D R. 1990, Watershed Delineation with Triangle-Based Terrain Models. Journal of Hydraulic Engineering, 116(6):1232–1251.

Jones, C. B., and Abraham, I. M., 1987. Line Generalisation in a Global Cartographic Database. Cartographica, 24, 3, 32–45.

Li Z. and Openshaw S., 1992, Algorithm for automatic line simplification based on a natural principle of objective generalization, International Journal of Geographic Information Systems, 6(5):373–389

Liu Y. and Zhai J.S., 2005, An automatic method for recognizing topographic terrain lines, Hydrographical Surveying and Mapping,25(3):21–23

Mandelbrot, B. B., 1967. How long is the Coast of Britain? Statistical Self-similarity and Fractional Dimension. Science. 156, 3775, 636–638.

Mandelbrot, B. B., 1975. Stochastic models for the Earth's relief, the shape and the fractal dimension of the coastlines, and the number-area rule for islands. Proc. Nat. Acad. Sci. (USA), 72,10, 3825–3828.

Mandelbrot, B. B., 1982. The Fractal Geometry of Nature. W. H. Freeman & Co.

Mandelbrot, B. B., 1985. Self-affine Fractals and Fractal Dimension. Physica Scripta, 32, 257–260.

Mark D M. 1984, Automated Detection of Drainage Network from Digital Elevation Model. Cartographica, 21(3) :168–178.

Mark, D. M. and Aronson, P. B., 1984. Scale-Dependent Fractal Dimensions of Topographic Surfaces: An Empirical Investigation, with Applications in Geomorphology and Computer Mapping. Mathematical Geology, 16(7): 671–683.

McMaster, R. B. (1986): A Statistical Analysis of Mathematical Measures for Linear Simplification. The American Cartographer, Vol.13, No. 2, 103–116.

McMaster, R. B. (1987): Automated line generalization. Cartographica, Vol. 24, No. 2, 74–111.

McMaster, R. B., 1989. Introduction to Numerical Generalization in Cartography. Cartographica, 26,1, 1–6.

Nakos, B., 1990. Digital Representation of Cartographic Phenomena Based on Fractal Geometry Theory. Application on Terrain's Relief Using Digital Models. Phd. Thessis. Department of Rural and Surveying Engineering, National Technical University of Athens. (In Greek)

Nakos, B., 1996. The Use of Fractal Geometry Theory for Performing Digital Generalization Tasks. Proceedings of the 2nd National Cartographic Conference, Hellenic Cartographic Society, 293–301. (In Greek)

Nakos, B., 1999. Comparison of Manual Versus Digital Line Simplification. Proceedings of 3rd Workshop on Automated Map Generalization. International Cartographic Association. Commission of Map Generalization. (http://www.survey.ntua.gr/main/labs/carto/Comparison.PDF)

O'Callaghan J F, Mark D M.1984, The Extraction of Drainage Networks from Digital Elevation Data. Computer Vision, Graphics and Image Processing, 28(3) :323–344.

Qian J, Ehrich R W, Campbell J B. 1990, An Expert System for Automatic Extraction of Drainage Networks from Digital Elevation Data. IEEE Transactions on Geoscience and Remote Sensing, 28(1): 29–45.

Ramer U., 1972, An Iterative Procedure for the Polygonal Approximation of Plane Curves. Computer Graphics and Image Processing, 1, 244–256.

Richardson, L. F., 1961. The problem of contiguity: an appendix of statistics of deadly quarrels. General Systems Yearbook, 6, 139–187.

Scheidegger A E.1996, Effect of Map Scale on Stream Orders. Bulletin International Association of Science Hydrology,11(1):56–61.

Shreve R L.1996, Statistical Law of Stream Numbers. Journal of Geology, 74(1): 17–37.

Strahler A N.1952, Dynamic Basis of Geomorphology. Bulletin of Geological Society of America,63 (7): 923–938.

Tang L. 1992,Automatic Extraction of Specific Geo-morphological Elements from Contours. in: Proceedings of the 5th SDH conference. Charleston , SC. 2:554–556.

Töpfer, F., and Pillewizer, W., 1966. The Principles of Selection. The Cartographic Journal, 3,1, 10–16.

Veregin, H. (1999): Line Simplification, Geometric Distortion, and Positional Error. Cartographica, Vol. 36, No. 1, 25–39.

Veregin, H. (2000): Quantifying positional error induced by line simplification. International Journal of Geographical Information Science, Vol. 14, No. 2, 113–130.

Visvalingam, M., and Wyatt, J. D., 1993. Line generalization by repeated elimination of points. The Cartographic Journal, 30, 1, 46–51.

Wang, Z., Muller, J.-C. (1998): Line Generalization Based on Analysis of Shape Characteristics. Cartography and Geographic Information Systems, Vol. 25, No. 1, 3–15.

Weibel R.1992,Model and Experiments for Adaptive Computer2Assisted Terrain Generalization, Cartography and Geographic Information Systems, 19(3):133–153.

Weibel, R., 1997. Generalization of Spatial Data: Principles and Selected Algorithms. In Algorithmic Foundations of Geographical Information Systems (eds. van Kreveld et al.) Springer-Verlag.

Werner C. 1988, Formal Analysis of Ridge and Channel Pattern in Maturely Eroded Terrain. Anneals of the Association of America Geographers, 78(2): 253–270.

Wolf G W. 1988, Weighted Surface Networks and Their Application to Cartographic Generalization,in: Proceedings of Visualization Technology and Algorithm, W Barth (ed.), Berlin : Springer.

Yan H.W. Chu Y.D. Yang S.W. and Sun J.G. 2007, Foundation of Principles and Algorithms in Computer-Aided Cartography, Beijing: Science Press.

Yoeli P.1984, Computer-Assisted Determination of the Valley and Ridge Lines of Digital Terrain Models, International Yearbook of Cartography, XXIV:197–206.

Zhang J. 2002, On the theory and implementation techniques of multi-resolution spatial data models, PhD dissertation, Wuhan: Postgraduate School of Chinese Academy of Science, China.

Zhang L.L. Wu F. and Wang L.H., 2005, Determination and applications of spatial relations among contour lines, Bulletin in Surveying and Mapping, 17(8):19–22

Zhu W.B. Bi R.T. Guo P.S. and Zhang J., 2008, Research on the method of the auto-generalization of contour based on the algorithm of Multi-resolution Simplification, Journal of Shanxi Agricultural University (Natural Science Edition), 28(3): 2–7

Chapter 4
Description and Generalization of Road Networks

4.1 Introduction to Road Networks

At least three types of features are displayed on large or intermediate scale (including 1:5K, 1:10K, 1:25K, 1:50K, 1:100K, 1:250K, and 1:500K) maps in the form of linear networks, i.e. roads, rivers and boundaries. However, they are different in structures and patterns. Roads are human-made structures, and they are generally arranged in order, intersect with each other, and form networks. A road network (also including street networks) geometrically consists of polygons and lines (Haggett 1967; Agrawala and Stolte 2001).

Networks are ubiquitous in nature and society, such as the Internet, networks of things, acquaintance networks, electric grid networks, and biologic networks (Newman 2003; Walter 2008). Among these, road network has been a central subject of network studies and has received intensive interdisciplinary attention during the past several years for two reasons (De Serres and Roy 1990; Yu 2001; Zhang 2004; Zhang and Li 2011; Weiss and Weibel 2013). The first is that road networks are common and easy to access spatial complex networks whose topology is strongly constrained by their spatial and geographic embeddings. The second is that road networks have a close relationship with human life and city involvement.

4.2 Description of Road Networks

The description of road networks has gained much attention for decades. At least four categories of methods for describing road networks can be found in the literature, i.e. graph theory-based methods, stroke-based methods, information theory-based methods, and measures.

H. Yan, *Description Approaches and Automated Generalization Algorithms for Groups of Map Objects*, https://doi.org/10.1007/978-981-13-3678-2_4

4.2.1 Graph Theory-Based Methods

4.2.1.1 Basic Concepts of Graph Theory

Besides traveling salespersons problem, circuit board design and minimum spanning networks, graph theory has also been applied to many other important problems including minimum vertex coloring, modelling Markov chains, enumeration of chemical molecules, network flows and scheduling critical path analysis. Its applications in geography are not new either, and many are familiar with the problem of the bridges of Konigsberg; the graph theory has also longly been used in quantitative geography and landscape ecology (Weibel and Dutton 1998).

This chapter focuses on the application of graph theory in network map generalization. It is easy to see generalization as a set of operators (simplify, omit, merge, etc.) whose application has been governed by controls (e.g. scale, aesthetics, content, theme), but generalization is not the mere analysis of spaces but the communication of spatial concepts. So, here applications of graph theory are discussed in the context of network generalization.

The following paragraphs present some basic definitions, concepts and examples of graph theory from the perspective of road networks description.

In mathematics, the graph theory is the study of graphs, which are used to model pairwise relations between objects with mathematical structures. A graph G = (V, E) in this context is made up of vertices, nodes, or points (i.e. V) which are connected by edges, arcs, or lines (i.e. E). A graph may be undirected, meaning that there is no distinction between the two vertices associated with each edge, or its edges may be directed from one vertex to another. Graphs are one of the prime objects of study in discrete mathematics.

For example, the graph in Fig. 4.1a consists of five vertices (A, B, C, D, E) and eight edges (AB, AD, CD, BC, DE, CE, AC, AE). The degree of a vertex is the number of edges who have that vertex as an endpoint (for example vertex D is of degree 3, having edges, AD, CD and DE). The degree for each vertex is shown

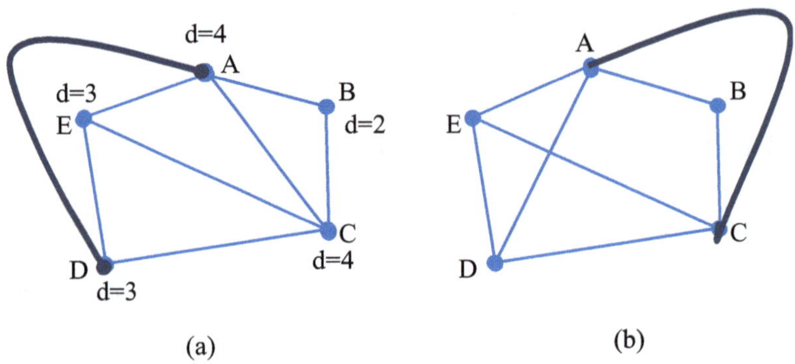

Fig. 4.1 Isomorphism of graphs. (**a**) A graph and the degrees of its nodes; and (**b**) an isomorphic graph of the graph in (**a**)

beside the vertex. Please note that (b) is isomorphic to (a), though the positions and drawings of the edges are different, the vertices have the same degree incident to one another (having only the same degree sequence does not guarantee isomorphism). Thus, a graph can appear visually different from the network from which it was derived; for example, in a graph derived from a road network, the crossings of edges do not necessarily represent overpasses.

In addition to the above basic concepts, the notions of complete graph, the complete bipartite graph and 他 the loop are also used in describing networks. Readers can know them from literatures (Hart et al. 1968; Mackaness and Beard 1993).

4.2.1.2 Describing Road Networks Using Graph Theory

As far as the description of road networks using graph theory is concerned, researchers in cartography and geographic information systems are interested in measuring traffic generated by nodes, flow along links, degrees of accessibility and connectivity, spatial extents of a network, network associations along a route, and influence of one place on other places on a route or in a network.

Graph theory reduces road networks to a mathematical matrix whereby an edge (i.e. link) is a line segment (for example, roads, rail lines, etc.) between locations. A vertex (i.e. node) is a location on the road network that is of interest (for example, towns, road intersections, etc.). Any graph for road analysis is and always has constraining boundaries.

Here, the four key concepts are defined (see Fig. 4.2).

Node: a location on a transportation route that has the capacity to generate traffic (flow).

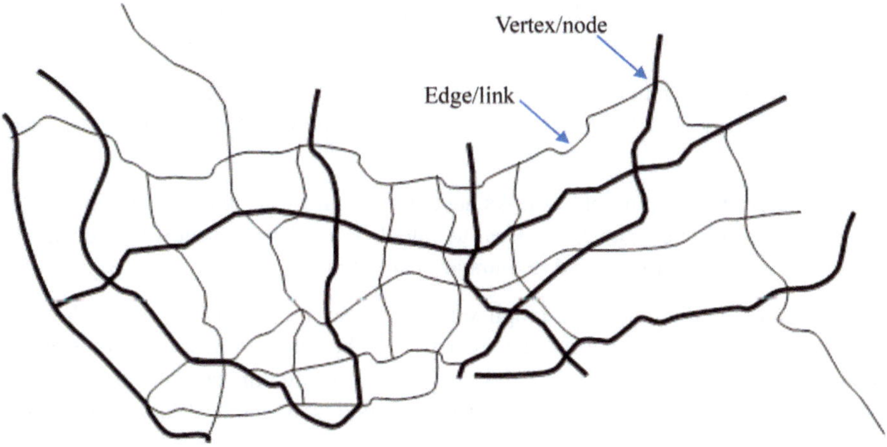

Fig. 4.2 A road network

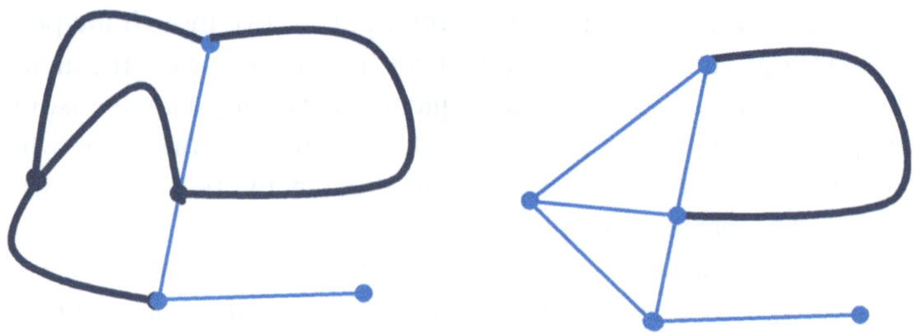

(a) (b)

Fig. 4.3 An example of planar graph. (**a**) Road system in real world; and (**b**) Plana graph of the road system

Link: the connection between two nodes along which flow occurs.
Route: a series of connected links.
Network: a system of nodes and links which consist of several modal types (road, rail, etc.), but is typically of a single mode type.

A road graph may be a directed graph whose direction of flow is explicit or an undirected graph that has no flow direction implied. It may contain loops (a loop flows from a node into itself). A road graph can be planar (Fig. 4.3) that means all links (edges) meet at nodes (vertices), or non-planar that means some links (edges) may cross each other.

The following paragraphs give a number of measures that are usually used in describing road networks.

1. Topologic distance: it is a unit-less measure where the distance between two nodes is coded as a single step. The distance is only implied (e.g. more steps = longer distance). The more nodes a route passes from the start node to the ending node, the longer the topological distance is. So, it is a substitution for real world distance that is not used here.

Figure 4.4 shows an example of the real-world distance (Fig. 4.4a) and topological distance (Fig. 4.4b). It also presents an example of topological distance with two steps.

2. Diameter: it is the number of links needed to connect the two most remote nodes in a network. The route that is used to calculate the diameter must be the shortest one. Backtracks, loops, and detours are excluded.

Actually, the diameter of a graph sometimes can be referred to as the 'longest, shortest path'. In other words, the diameter is the maximum number of steps to connect any two points on a graph. It is an index for evaluating complexity of graphs.

For example, in Fig. 4.5, the most distant nodes are node 3 and node 8. The shortest path from node 3 to node 8 is 3→2→7→8; so, the diameter of the graph is 3.

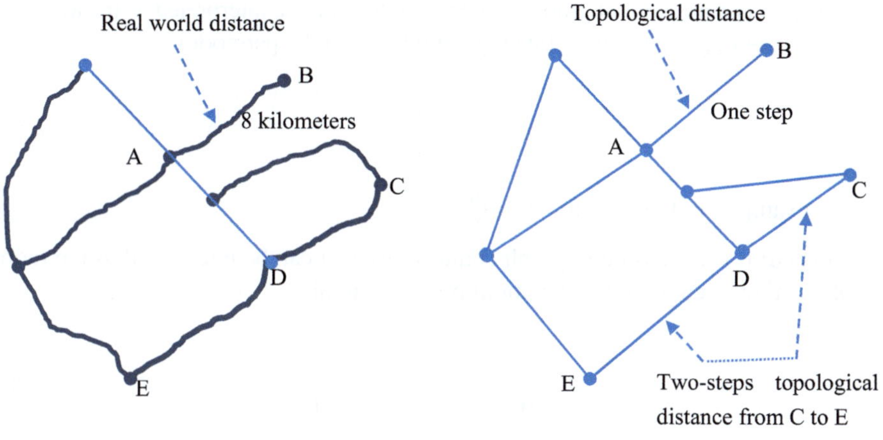

Fig. 4.4 Concept of topological distance in road networks. (**a**) Real world distance; and (**b**) topological distance

Fig. 4.5 Concept of the diameter of a graph

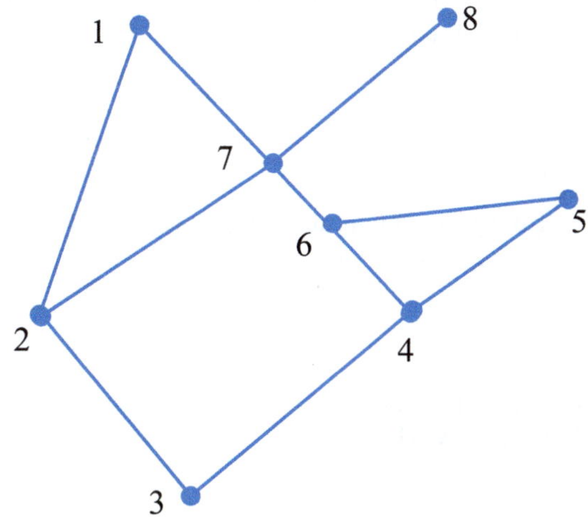

3. Gamma γ: it is a measure of graph connectivity and is a simple measure to use. It can be thought of as the percent existing routes to potential routes.

$$\gamma = \frac{e}{0.5v(v-1)} \tag{4.1}$$

where, v is the number of vertices in the graph, and e is the number of edges in the graph.

For example, in Fig. 4.5, e = 10 and v = 8. So, $\gamma = \frac{10}{0.5 \times 8(8-1)} = \frac{10}{28} = 0.357$.

4. Beta β: it is a measure of graph connectivity that can be interpreted as the average number of edges per vertex (average number of links per node).

$$\beta = \frac{e}{v} \tag{4.2}$$

For example, in Fig. 4.5, $\beta = \frac{e}{v} = \frac{10}{8} = 1.25$.

5. Alpha α: it is a measure of graph connectivity that can be interpreted as the ratio of existing circuits to the maximum possible circuits.

$$\alpha = \frac{e - (v - 1)}{0.5v(v - 1) - (v - 1)} \tag{4.3}$$

For example, in Fig. 4.5, $\alpha = \frac{10-(8-1)}{0.5\times 8(8-1)-(8-1)} = \frac{3}{21} = 0.143$.

6. Pi index π: it is an index for the relationship between the total length of the graph L and the distance along its diameter D. A high index shows a developed network. It is a measure of distance per unit of diameter and an indicator of the shape of the network.

$$\pi = \frac{L}{D} \tag{4.4}$$

Taking Fig. 4.6 as an example, L = 8+7+4+6+4+5+10+7+7+8 = 66, and D = 7 +7+8 = 22. So, $\pi = \frac{66}{22} = 3$.

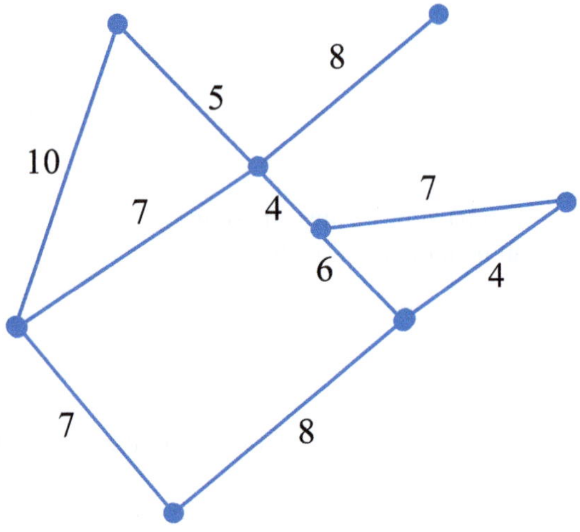

Fig. 4.6 Calculation of the diameter of a graph

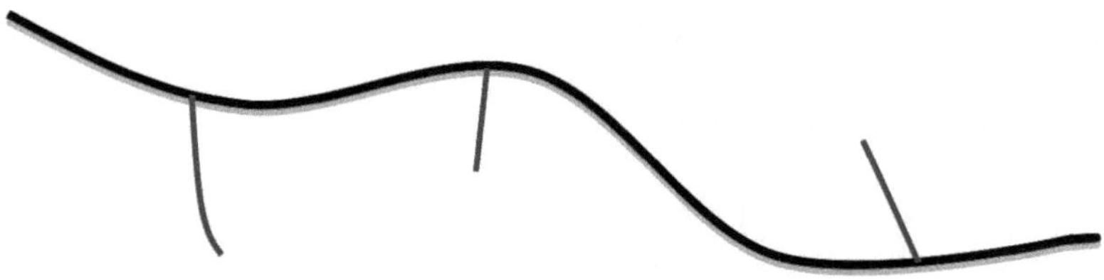

Fig. 4.7 Perceptual grouping of roads: good continuation takes effect

7. Eta η: it is the average length per link. Adding new nodes will cause a decrease in η as the average length per link declines.

$$\eta = \frac{L}{e} \tag{4.5}$$

In Fig. 4.6, $\eta = \frac{66}{10} = 6.6$.

8. Detour Index DI: it is a measure of the efficiency of a network in terms of how well it overcomes distances or the friction of distances. The closer the detour index gets to 1, the more the network is spatially efficient. Networks having a detour index of 1 are rarely, if ever, seen and most networks would fit on an asymptotic curve getting close to 1, but never reaching it.

$$\mathrm{DI} = \frac{DD}{TD} \tag{4.6}$$

For example, in Fig. 4.7, the straight distance (DD) between two nodes may be 40 km but the transport distance (TD, i.e. real distance) is 50 km. The detour index is thus 40/50 = 0.8.

9. Network Density ND: it measures the territorial occupation of a transport network generally in terms of kilometer of links (L) per square kilometers of surface (S). The higher it is, the more a network is developed.

$$\mathrm{ND} = \frac{\mathrm{L}}{\mathrm{S}} \tag{4.7}$$

For example, the area of a city is 1000 Km^2, and the total length of the roads in the city is 245 Km. Then, the network density of the city is $\mathrm{ND} = \frac{245\ \mathrm{Km}}{1000\ \mathrm{K}m^2} = 0.245\left(\frac{\mathrm{Km}}{\mathrm{K}m^2}\right)$.

4.2.2 *Stroke-Based Methods*

4.2.2.1 Origination of Strokes

The human visual system can spontaneously organize elements of the visual field, even with no high level or semantic knowledge. The Gestalt psychologists have studied these phenomena and they have discovered how some arrangements of picture elements tend to be regarded as 'belonging together' (Ren et al. 2015), forming natural groups. Many perceptual grouping principles have been identified, such as proximity, similarity, good continuation, symmetry, closure, parallelism, collinearity, co-termination and continuity. The road network in Fig. 4.8 shows the importance of good continuation in road network perception and organization: people usually perceive this network consisting of the longest road with the other three shorter roads incident on it, rather than as seven connected roads.

The principle of good continuation can be expected to be the dominant principle, but if the thematic information is available then the principle of similarity (of thematic attributes) can also play an important role. The perceptually salient features of the network can therefore be identified by applying the same principles to group (i.e. concatenate) the arcs into chains. These chains, which are termed 'strokes', do not branch, although they may cross (unlike their constituent arcs). The term 'stroke' is prompted by the idea of a curvilinear segment that can be drawn in one smooth movement and without a dramatic change in style. As far as road networks are concerned, a stroke usually is a chain of road segments with continuous curvature, i.e. with "good continuation" (Thomson and Richardson 1999; Thomson and Brooks 2000).

The concept of stoke has been widely used in many aspects after its birth (Thomson and Richardson 1999; Thomson 2006). For example, strokes can be used for pattern detection (Ester et al. 1996; Heinzle et al. 2005; Thom 2005; Harrie

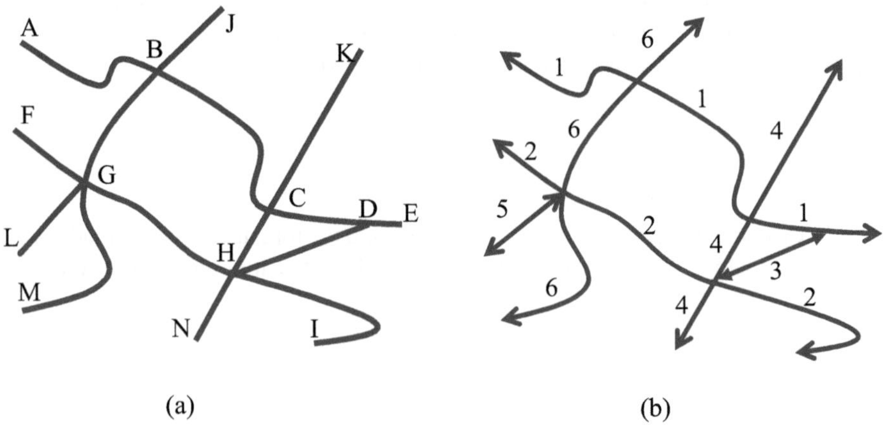

Fig. 4.8 Strokes are organized only by the principle of good continuation

and Weibel 2007; Heinzle and Anders 2007; Yang et al. 2011), topological analysis (Touya 2010) or hierarchical modeling of roads (Jiang 2004, 2009; Tomko et al. 2008; Jiang 2009; Jiang 2009). Of course, most often, strokes can be used for the selection of road networks (Thomson and Richardson 1999; Edwardes and Mackaness 2000; Thomson and Brooks 2000; Chaudhry and Mackaness 2005; Heinzle et al. 2005; Liu et al. 2010; Yang et al. 2011; Li and Zhou 2012).

4.2.2.2 Basic Rules/Factors Considered in Forming Strokes

At least two factors take effect in forming strokes, i.e. good continuation and thematic attributes of roads.

Figure 4.8 provides an example of a network and the set of constituent strokes found by such an analysis. Figure 4.8a presents a simple road network, with no thematic attributes. This network, if it is considered as a graph, consists of arcs/edges and nodes. Figure 4.8b shows the six strokes that were identified in the network; these have been labelled 1 through 6, and their endpoints marked with arrows. The values of the numbers used as labels have no special significance. The description of the strokes formed in Fig. 4.8b is listed in detail in Table 4.1

The road network shown in Fig. 4.9 is the same in structure as that shown in Fig. 4.8, but with additional thematic information: three categories of roads are now represented. In such a situation, the process of arc concatenation to organize the strokes is further constrained by imposing limits on the compatibility of arc thematic categories. In this example, it will be assumed that (1) the adjacent roads belonging to the same category are easy to be perceived as a stroke; and (2) it is allowable to concatenate an arc of category 1 with an arc of category 2, and similarly to arcs of categories 2 and 3, but those arcs of categories 1 and 3 cannot abut in a stroke.

In Fig. 4.9, road segment JBG concatenates GL but not GM and forms a stroke, because they belong to the same category (category 2). Road segments AB, BC and CE can form a stroke, because AB belongs to category 3, and BC belongs to category 2, and CE belongs to category 1. On the contrary, road segment FGH and road segment HI does not go to the same stroke, because the former belongs to category 3, and the later belongs to category 1.

Table 4.1 Description of the strokes found in Fig. 4.8b

Stroke	First terminal node	Second terminal node	#Arc/#edge
1	A	E	4
2	F	I	3
3	D	H	1
4	K	N	3
5	G	L	1
6	J	M	3

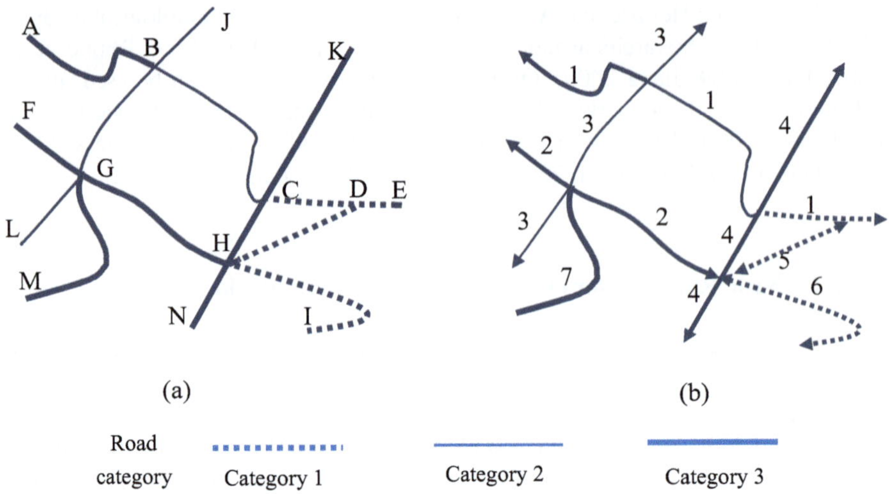

Fig. 4.9 Strokes are organized considering the principle of good continuation and thematic attributes

Fig. 4.10 Strokes are scale-dependence. (**a**) Map scale is 1:1 K, and road segments AB, BC and CE form a long stroke; and (**b**) map scale is 1:10 K, and road segments AB, BC and CD form a long stroke. The two maps are not shown to the true scale

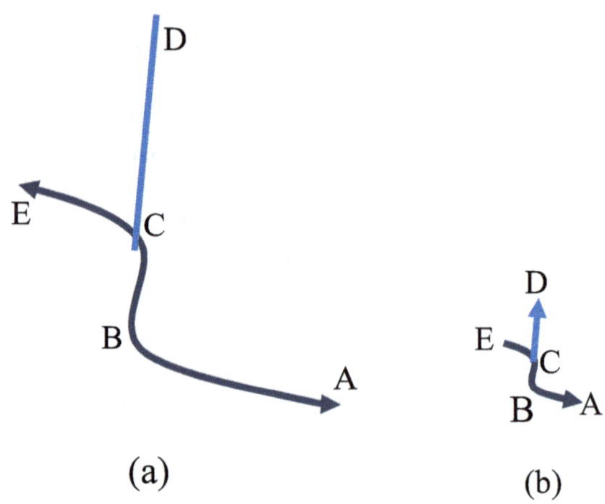

4.2.2.3 Scale Dependence of Strokes

Figure 4.10 represents a portion of a road network where a minor road joins a major road on a bend. Figure 4.10a shows a larger scale view of the road, and it is easy to see that road CD intersects with the other long smooth road, and therefore the latter itself is a stroke. When this map at scale 1:1 K is simplified and get a map at smaller scale 1:10 K (Fig. 4.10b), the short road segment of AC on Fig. 4.10a is deleted due to its illegibility on the smaller scale map, and road segments BC and CD become a nearly straight line and can be concatenated visually, and therefor AB, BC and CD

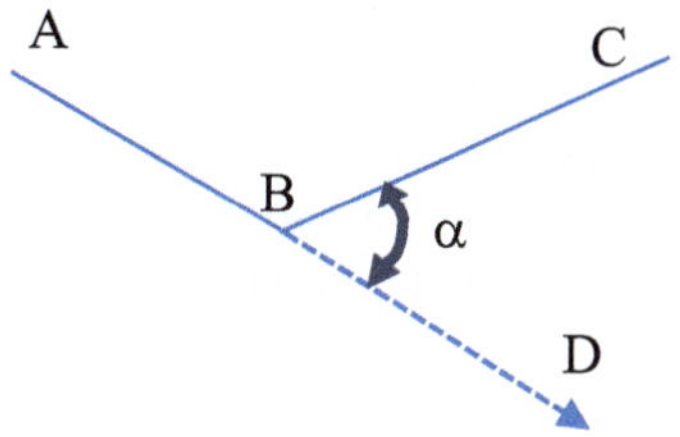

Fig. 4.11 α is the deflection angle between road segments AB and BC when the road run direction is ABC

form a long stroke. This example presents that strokes of a road network may change with the change of map scale. In other words, strokes are scale-dependence.

4.2.2.4 Formation of Strokes

Approaches to concatenating road segments into strokes can be classified into three types: geometric, thematic, and hybrid approaches (Zhou and Li 2012), and these approaches abide by a number of criteria and strategies. Here, concatenation criteria are used to determine whether two adjacent segments can be linked to form a stroke or a part of a stroke. The concatenation strategies are used to determine how each segment is linked with its neighbours.

By far three concatenation criteria (i.e. geometric criterion, thematic criterion and hybrid criterion) and three strategies (i.e. every-best-fit, self-best-fit and self-fit) for building strokes have been proposed. To introduce the approaches easily, the criteria and the strategies are presented before the approaches.

The three criteria may be summarized as follows:

Geometric criterion- Two road segments with less deflection angle (see Fig. 4.11 an example of deflection angle) at its run direction can be easily perceived as following the principle of good continuity, and thus, they are tended to be concatenated to form a stroke than other ones.

Thematic criterion- Two road segments with the same name, or are of the same class, or are of two similar classes, can be more easily to be concatenated to form a stroke.

Hybrid criterion- A combination of the geometric criterion and the thematic criterion. Under this condition, thematic criterion is usually considered first, followed by the geometric criterion. For example, when a road network is given for building strokes, road class/category is considered first, which means only the roads of the same or similar class/category are qualified to be linked to form a stroke, and then the geometric criterion, such as deflection angle, may be used to determine the best concatenation.

The three strategies are as follows:

Every-best-fit- Compare every pair of road segments and select the best-fit pair for concatenation.

Self-best-fit- Select a segment that is the best-fit for a starting segment, regardless of whether the selected segment has a better choice.

Self-fit- Randomly select a segment for a starting segment as long as the concatenation of the starting segment with another segment satisfies the adopted concatenation criterion.

The Geometric Approach

The geometric approach is based on the original idea of building road segments into strokes proposed by Thomson and Richardson (1999). The thought is very simple but natural: 'elements that appear to follow in the same direction tend to be grouped together' (Thomson and Brooks 2000), however, it follows the principle of good continuation that is widely accepted in visual perception.

In the process of stroke construction, a simple criterion is employed to judge which two adjacent road segments can be linked to form a stroke or a part of the stroke. This geometric criterion is: the less the deflection angle, the more possible the road segment is concatenated. It is obvious that a less deflection angle value means that a straighter line results from the concatenation. If the deflection angle is 0, the two concatenated roads form a straight line. Otherwise, if the deflection angle between the two road segments is great (e.g. greater than 90°) it is generally not appropriate to concatenate them, because the result of their concatenation does not appear to follow the principle of good continuity (going in the same direction). Therefore, it has been suggested to make use of a small deflection angle between 40 and 60° (Chaudhry and Mackaness 2005) as the threshold to ensure that all the strokes can follow the principle of good continuity.

Suppose that a road network is given, strokes can be organized starting from an arbitrary road segment (taking Fig. 4.12 as an example, AB is the starting road segment). Run from the beginning endpoint (it is A in this example) to the ending endpoint (it is B in this example) of this road. If the ending endpoint of this road is only connected with one other road (say, road b), take the ending endpoint as the beginning endpoint of road b, and repeat this process. Otherwise, if the ending endpoint is connected with at least two other road segments, it is necessary to decide which one to connect with to form a stroke. Under this condition, the above three strategies should be considered, i.e. self-fit, self-best-fit, and every-best-fit.

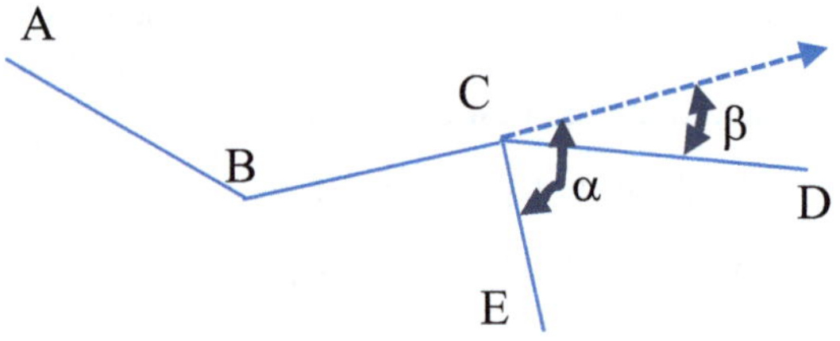

Fig. 4.12 Concatenating strokes using the geometric approach: road ABCD is the stroke

Both the self-fit and self-best-fit strategies can be called 'selfish.' They treat the starting road segment as the first priority and consider only the possible concatenations from the starting road segment. Thus, different results may be obtained if one starts from different road segments. By contrast, an optimum result may be obtained by every-best-fit, as this strategy considers all the possible concatenations at each intersection.

The Thematic Approach

Roads exist in the geographic space; thus, they should not be viewed as simple geometric lines. Besides the geometric characteristics, their thematic attributes should also be taken into accounts in building strokes. It is an easy way to make use of road/street names (Jiang and Claramunt 2004) as a constraint to construct strokes. To be more concrete, those segments with the same names should be linked to form a stroke. Nevertheless, only considering road/street names puts the thematic approach into trouble:

1. As far as a road network saved in a database is concerned, some roads do not have names; some road names might be missing.
2. It is a common phenomenon that a long winding or/and zigzag road/street might be split into a couple of sub-roads and each sub-road is given a name. For example, the long sinuous Queens Road in a city may be broken into Queens Road East, Queens Road Central, and Queens Road West. In this case, it is difficult to judge if the three sub-roads are the same road.
3. Very rarely but it exists in the geographic space in our life that two or more road segments meeting at an intersection have the same name.

In addition to road names, other attributes, such as class, road type, road width, number of lanes, and traffic speed, can also be used in organizing stroke. Which attributes should be taken into consideration usually depends on target map scale and map readers.

The Hybrid Approach

The reason for combining the geometric approach and the thematic approach to generate a hybrid approach is that: through analysis of many road databases, it has been found that although road classes and road names are usually recorded on road segments, other attributes are generally unavailable. So, only thematic information sometimes is not enough for decision making in building strokes, and an assistant from geometric information becomes necessary. This leads to the hybrid approach, i.e. an integration of the thematic approach and the geometric approach.

A key issue that should be firstly considered in the hybrid approach is the priority of the two approaches. In other word, is the geometric approach or the thematic approach executed first in building strokes? In practice, either of the two can be at the first position. They may be named "thematic-geometric" strategy and "geometric-thematic" strategy. As far as the former is concerned, thematic attributes are used as the primary criterion and geometric elements are used as supplementary one. As far as latter is concerned, geometric elements are used as the primary criterion and thematic attributes are used as supplementary one. For any geometric criterion, a rule that should always abide by is: two road segments can be finally concatenated only when they have the identical thematic attribute.

4.2.3 Information Theory-Based Methods

Because maps are carriers of geographic information, the important information containing in the larger scale maps should be transmitted to the smaller scale maps after map generalization. In this sense, it is necessary to know what kinds of information road maps contain, and what information is more important, than other information and therefore should be kept in smaller scale maps after the larger scale maps are generalized.

Existing achievements have classified the information containing in road networks into four categories: statistical, metric, topological, and thematic (Richardson and Thomson 1996; Yan and Weibel 2008). On the other hand, previous study has revealed that generalization of road networks can be analyzed and implemented at three different spatial levels, i.e. macro, mezzo and micro scale generalization (Liu et al. 2010).

Macro-scale generalization- it focuses on the global and general information of the will-be generalized road networks. This information includes the total number of the roads, the total length of the road network, the total area the road network occupies, etc.

Mezzo-scale generalization- it pays attention to obtain various spatial relations among the roads, and/or among the roads and the facilities and features related to the road network.

Micro-level generalization: it allows for the analysis of geometric and semantic property of single road segment, such as the road segment frequency of usage, type, width and length.

If the four types of information are considered in the three levels of road network generalization, 4×3 grids may be drawn with the columns corresponding to the types of information and the rows corresponding to the levels of road network generalization (See Fig. 4.13 as an example).

By this way, the information containing in road networks can be further classified into sub-classes and each sub-class corresponds to a cell. For example, thematic information at the micro level is marked in the cell of the second row and the third column. One important advantage of this description is that: the types of information considered by each algorithm or a class of algorithms for road generalization can be clearly represented by the cells. Furtherly, comparisons of the types of information

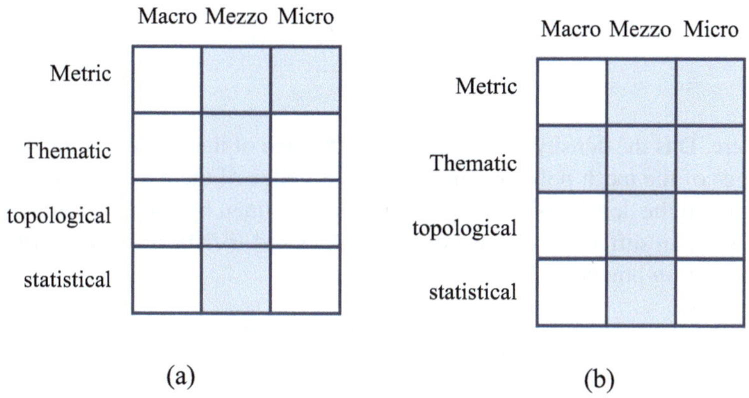

Fig. 4.13 Types of information that might be considered in different levels of map generalization algorithms. (**a**) A stroke-based algorithm (Li and Dong 2010), and (**b**) a graph-based algorithm (Mackaness and Beard 1993)

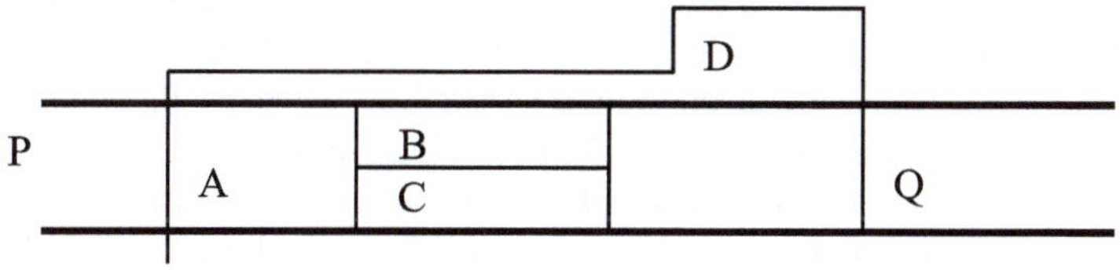

Fig. 4.14 Meshes of a road network

considered or not considered by different algorithms can be explicitly expressed using the 4 × 3 grid. For example, Fig. 4.13 presents the types of information considered by a graph-based algorithm and a stroke-based algorithm.

4.2.4 *Mesh-Based Method*

4.2.4.1 Definition of Meshes of Road Networks

If all edges of a simple polygon are the roads of a road network, this polygon is a mesh of the road network. For example, Fig. 4.14 is a road map. A, B, C, and D are meshes of the road network, while P and Q are not. Here, a simple polygon means the polygon has no other polygons in it, and it has no self-intersected edges.

4.2.4.2 Density of Meshes

The density of a mesh is defined as the ratio of the total length of the edges of the mesh and the area of the mesh (Benz and Weibel 2013). This can be expressed by Formula 4.8.

$$D = \frac{\sum_{i=1}^{n} L_i}{A} \tag{4.8}$$

Where, D is the density of the mesh; A is the area of the mesh; n is the number of the edges of the mesh polygon; and L_i is the *ith* edge of the mesh polygon.

Although the density of a mesh has also been defined by some researchers (Chen et al. 2009) in different way in Formulae (4.9) and (4.10), this formula has been seldom used in practice.

$$\lim D = \lim_{\varepsilon \to 0} \frac{P + L + \Delta}{A + A_{\varepsilon}} = \frac{P + L}{A} \tag{4.9}$$

$$D = \frac{P + L}{A} \tag{4.10}$$

Where, D is the density of the mesh; P is the total length of the edges of the mesh; A is the area of the mesh; $A_?$ is the area of the small region ?; Δ is the total length of the roads within ?; and L is the total length of the roads falling in the mesh.

To simplify the following discussion, if a road falls in a mesh, the road is called an overhanging road of the mesh (Fig. 4.15). In calculation of the density of a mesh, the total length of its overhanging roads in the mesh (i.e. L in Formulae (4.9) and (4.10)) are usually not considered in calculating the density of the mesh, due to the following reasons: the overhanging roads have nothing to do with the connectivity of the road network, their retaining or deletion in map generalization only depends on its geometric and semantic attributes, and they are generally deleted before other roads.

4.2.4.3 Topological Descriptions of Meshes of Road Networks

There are three categories of topological relations of road networks: relations among meshes, relations among roads, and relations among meshes and roads.

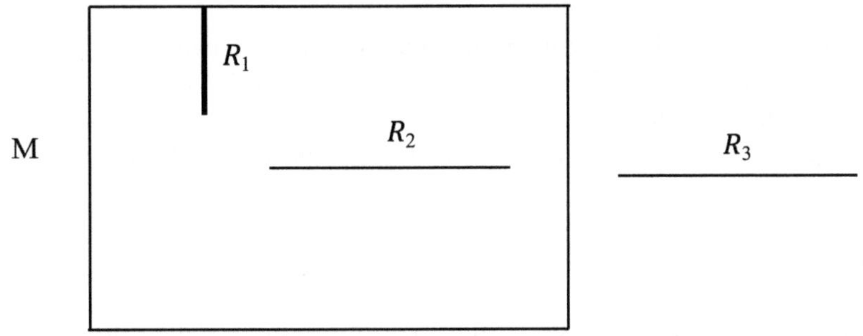

Fig. 4.15 Overhanging roads of the mesh M: R_1 and R_2 are overhanging roads of M, while R_3 is not

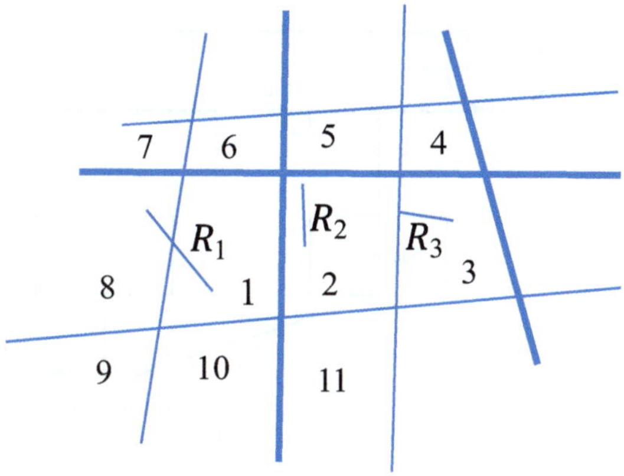

Fig. 4.16 Topological relations among meshes and among meshes and roads

Topological relations among meshes of road networks include neighbouring relations and separation relations.

Topological relations among meshes and roads (or road segments) include joint, disjoint, intersection, and containment relations.

Topological relations among roads (or road segments) include neighbouring relations, separation relations, and intersection relations.

A number of examples of the topological relations are shown in Fig. 4.16:

- Mesh 1 has four neighbors, and they are mesh 10, mesh 2, mesh 6 and mesh 8. Mesh 1 is separated from mesh 3.
- Mesh 2 contains road R_2, and mesh 3 contains road R_3.
- Road R_1 intersects with mesh 1.

More importantly, topological relations among meshes and roads should be expressed in an efficient data structure so that they can be used in road map generalization. A commonly used data structure might be defined as follows in C++:

```
Class CMesh
{
public:
  int ID;       //ID of the mesh
  int Nedge;     //the number of the edges of the mesh
  int *Edges;     //the array for storing the IDs of the edges of the mesh
  int *NMeshes;   //the array for storing the IDs of the neighouring
meshes of the mesh
  int *CRoads;    //the array for storing the IDs of the roads wholly
contained by the mesh
  int *IRoads;   //the array for storing the IDs of the roads intersecting
with the edges of the mesh
};
```

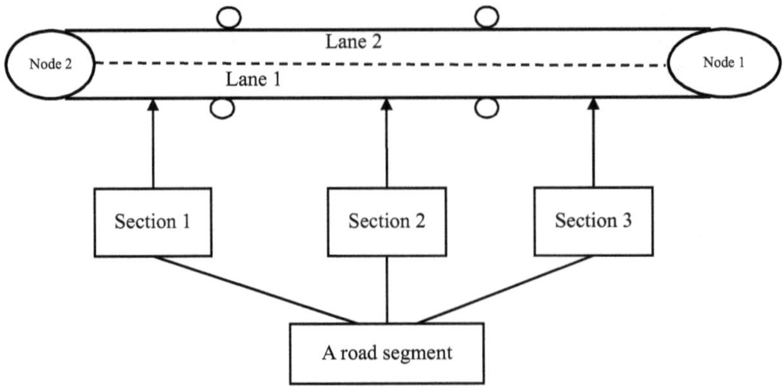

Fig. 4.17 Detailed description of a road segments

Of course, direction relations, distance relations and similarity relations among meshes are also important in describing road networks. However, they are simpler to calculate than topological relations, therefore they are not further discussed here.

4.2.5 Simulation-Oriented Road Descriptions

Roads are carriers of vehicles, such as buses, trucks and cars; thus, it is usually required to present reality or nearly reality models of roads to users (e.g. car drivers). In this sense, only topological and geometric information of road networks is necessary but not sufficient in constructing modern navigation systems for drivers and navigation control centres, because it does not contain adequately detailed geometric information that is mandatory in simulating reality of roads or road networks (Morisset and Ruas 1997).

Lanes (e.g. fast lane, car lane, truck lane, etc.) of roads are basic units for carrying vehicles; therefore, it is appropriate to describe lanes of roads in detail for the purpose of providing enough information for drivers and navigation control centres.

Figure 4.17 shows a description of a road segment. It is divided into three sections, and each section has its lanes and nodes. This is a commonly used description method in road simulation.

4.3 Fundamental Principles in Road Network Generalization

The generalization of a road network or a road map should be taken into accounts the functions and regional geographical characteristics of the road network or the road map. In addition, the original and target map scales and the purpose of the target map should also be considered. To achieve these objectives, many principles need to be abided by in road network generalization.

- Try to not damage the connectivity among roads and among roads and other relevant features in map generalization process.

Roads usually exist among/between buildings (e.g. settlements, scenic spots, man-made constructions etc.), therefore it is pertinent to consider the connectivity of the will-be-generalized road networks along with the generalization of the relevant buildings. If a road is deleted while the buildings at the starting and ending points are still retained but not connected on the target map, the deletion of the road is a bad operation and thus should be avoided.

- Essential geometric characteristics of the road network should be retained after map generalization.

Road networks usually have different plan patterns in different regions/cities. These patterns or characteristics should be kept on the target map, which can be evaluated by the similarity degree between the original and target road networks.

- Contrast relations of road density among different areas should be kept.

This means if an area on the original map has higher road density than another area, this area should still have higher road density on the generalized map.

- The more important a road is, the more probable it can be retained after road map generalization.

The degrees of importance of roads are generally complicated to calculate, because they are related to many factors such as road class, road quality, the number of vehicles on the road per hour. So, different factors should be selected, and different weights might be assigned to the factors considering different purposes of generalized maps.

- Although it is inevitably to damage the topological relations among roads in the process of road networks generalization, try to reduce the damage as less as possible. The less the damage of the topological information is, the more the information containing in the original road map is transmitted to the generalized road map.

4.4 Approaches for Generalizing Road Networks

Road network generalization has been a hot issue in the communities of geographic information systems and cartography for decades, and many achievements have been made. It can be found in literature, these approaches can be classified into five categories: graph theory-based, stroke-based, cartographic information theory-based, mesh density-based. The following paragraphs will select some typical algorithms/schemes for each category of the five approaches and discuss them in detail one by one.

4.4.1 A Graph Theory-Based Approach for Simplifying Road Junctions

Map generalization is required if geographic information at large scale should be represented at a smaller scale. The overall objective of map generalization is the meaningful grouping and symbolization of geographic phenomena that are appropriate for the target scale and intended map use. When road networks are generalized a number of objectives govern how road information is presented: (1) the portrayal of generalized road maps should be as accurate as possible; and (2) the result maps convey the connectivity, structures and patterns of the original networks correctly. When cartographers' work is observed, it is apparent that a variety of techniques are used to generalized the road information such as line simplification, line removal and junction simplification. These techniques usually generate different results depending on their sequences and degrees of applications. Automated map generalization is a mimic of human cartographers' work. In this sense, the road network generalization can be viewed as the combination of two processes: (1) the reduction of connectivity between junctions (requiring the removal of sections of road), and (2) the reduction of complexity within the junction (requiring the simplification of the junction itself). Here addresses an approach (Mackaness and Mackechnie 1999) for simplifying road junctions using the graph theory.

Because the objectives of automated generalization are the same as for the manual approach, the information requirements for automated solutions are similar to those of the cartographers. To be specific, a map generalization system should know (1) when to generalize (where are dense or cluttered regions in the map), (2) how to generalize, (3) how to retain the intrinsic qualities of the phenomenon that are represented on the map, and (4) how to improve the legibility and clarity of the features on the map.

The characteristics of road networks can be viewed in two parts: (1) the internal metric, topological and attribute information that collectively convey the idea of a road network, and (2) the external associations and interactions with other geographic objects such as the intersection of a road and a river requiring a bridge, or the close physical association between roads and buildings. Thus, it is natural to identify dense regions (when, internal) using cluster analysis, and to find mechanism for generalizing junctions (how, internal and external) using the graph theory whilst retaining the connections and disconnections among a set of roads.

The following approach includes three steps: analysis, synthesis and evaluation. The first step is the cartometric analysis by which the various defining characteristic of the network are analyzed. The second step, synthesis is the creation of candidate solutions. The final step is the evaluation of the solution to assess whether additional

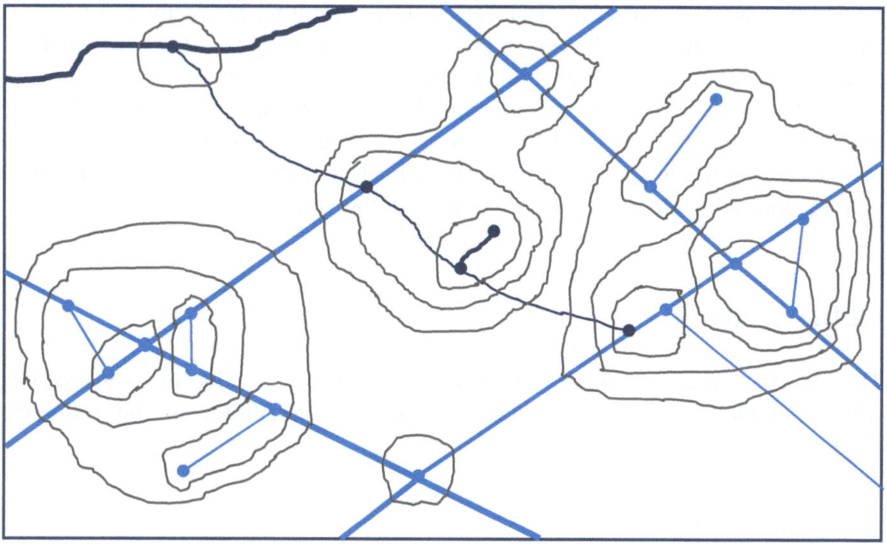

Fig. 4.18 Hierarchical clustering of the vertices of the road network

generalization is required. This step also includes evaluation of the impact on 'external' associations. For example, the simplification of a road may require movement, re-orientation or removal of buildings adjacent to the road. Thus, the generalization of road networks may trigger 'reactive' generalization operators on other groups of objects. The complications associated with such 'knock on' effects are outside the scope of this section.

A map with simulated data is used in the following discussion. The road junctions (including general vertices, pendant vertices and articulation vertices), road segments and cul de sacs are discernible on the map (see Fig. 4.18).

4.4.1.1 Analysis: Hierarchical Grouping of the Vertices

The purpose of this step is the detection of relatively dense regions of vertices (this being indicative of junctions). In this step, the topological structure is ignored, and the vertices are treated as simple points on a 2D plane. This is done by clustering the coordinates of the vertices of the road network. The grouping method consists of the following procedures:

1. The closest pair of vertices are grouped, and the position of their centroid is recorded.
2. The next closest pair of vertices/centroids are grouped, and the position of their centroid is recorded.
3. Continue with the procedure (2) until all vertices belong to one or other groups.

A dendrogram is a useful visualization of the resulting hierarchy and provides clustering information at all levels of granularity. By altering the 'depth' at which we choose to 'snip' the dendrogram branches and create clusters, it is possible to control the level of generalization of the junction. The hierarchical structure implicit in the dendrogram also provides an effective basis for efficiently compartmentalizing the map according to the groupings of vertices. Each 'branch' is visited in turn, and a graph is formed that represents the topology of the vertices (Fig. 4.18).

4.4.1.2 Synthesis

The objective of this step is to determine the importance value of each junction, which includes the four operations discussed by examples in the following paragraphs:

- Create graph,
- Contract graph,
- Determine new location of centroid, and
- Update vertex information.

Figure 4.19 shows a hypothetic road network. The dot circles represent the clustering process of the junctions. Here, the vertices within the clustered groups are numbered, and those vertices outside the groups are lettered.

Figure 4.20 is the dendrogram showing the spatial adjacency relations of the vertices in the road network of Figs. 4.19 and 4.21 shows a graph theoretic representation of the vertices. It is apparent from the graph in Fig. 4.20 that there

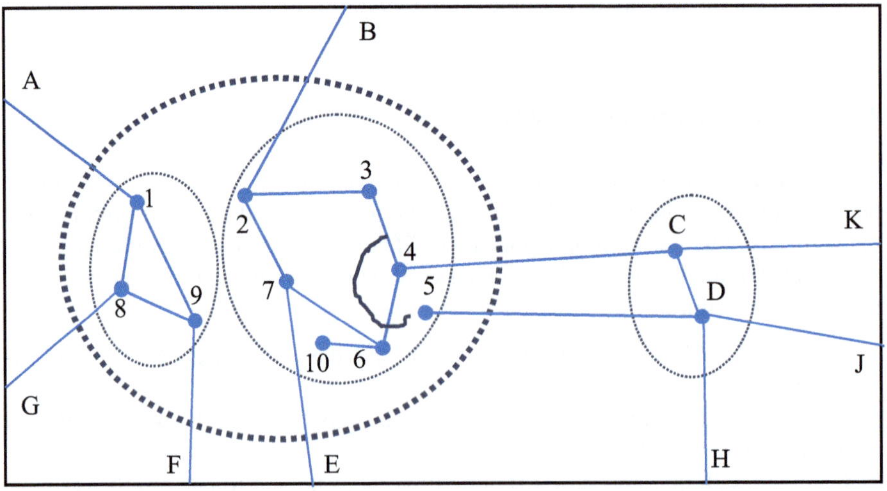

Fig. 4.19 Vertices clustering of the road network

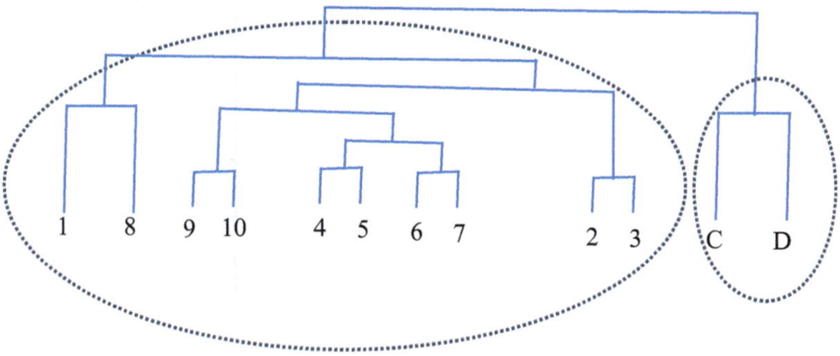

Fig. 4.20 A dendrogram of the clustered vertices

Fig. 4.21 A graph theoretical expression of the clustered vertices

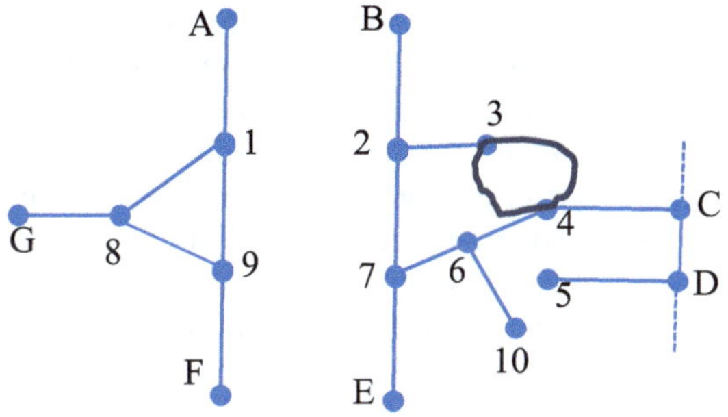

are two disconnected graphs in the road network; therefore, this quality should be retained after road network generalization.

Taking Fig. 4.19 as an example, the idea of the algorithm can be explained as follows.

The cluster in Fig. 4.19 consists of three subgraphs {1,8,9}, {2,3,4,6,7,10}, {5}. Those subgraphs with single edge (such as {5}) are omitted from further generalization, since they are end points of lines and are not reducible. For the remaining subgraphs, each vertex in the subgraph is examined in turn, and the vertices to which it is connected are identified. For 1, 8, 9, the connections are {1}, A, 8, 9, and {8} G, 9, 1, and {9} F, 8, 1. Any vertices connected solely within the subgraph are removed from the list. In this instance, none of the vertices is internal to the subgraph. Next, any repeated items are removed from these lists. What remains are 1, 8, 9, A, G, F. The centroid X is now calculated as the mean of the vertices in the subgraph (1, 8, 9). Lines are now drawn from each of the vertices (1, 8, 9) to the new centroid X (see Fig. 4.22) which now becomes the vertex common to the edges (AX, GX, FX). The process is repeated for the other graph. The vertices connected solely within the subgraph are removed (Vertices 3, 6, 10). Repeated items are removed from the list and what remains are (2, 4, 7, B, E, C) (see Fig. 4.22). Lines are now drawn from each of the vertices of the subgraph (2, 4, 7) to the new centroid Y which now becomes the vertex common to edges (BY, CY, EY). The result for both subgraphs is shown in Fig. 4.22.

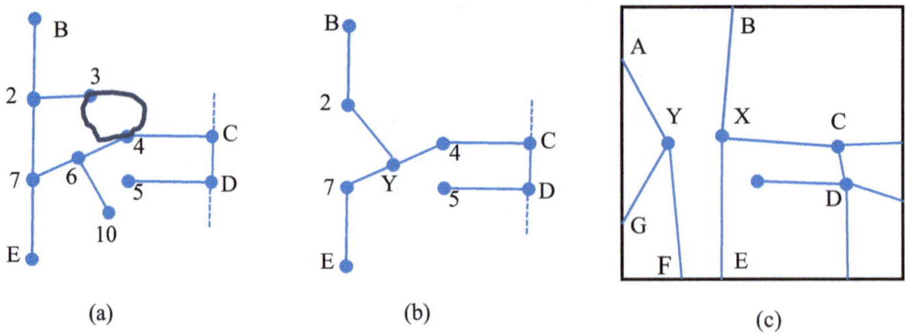

Fig. 4.22 Generalization process of the subgraphs and the resulting map

The last step is to update information pertaining to the new vertices. Though the degree of vertex Y is now 3, it represents a set of adjacent vertices whose cumulative degree is 17 (the sum of the degree of vertices 2, 3, 4, 6, 7, 10). This information can be used to reflect the degree of generalization of a junction. For example, Y could be symbolized in proportion to the number of contracted edges contained within the vertex. In addition to the removal of the loop between vertices 3 and 4, it should be noted that the algorithm has removed the short cul de sac. A cul de sac is an example of a pendant edge with a vertex of degree one. If it was a requirement that edges of relatively short distance be retained then it would be simple to formulate a rule that prevented these from being removed from the list. The information that defines cul de sacs can be gathered by searching for vertices in the subgraph with degree one and adjacent edges of a specified length (such as vertex 10 in Fig. 4.21).

4.4.1.3 Evaluation

After completing road network generation, the quality of the result is checked in this step, which includes evaluation of road network connectivities, topological relations among roads and other features such as settlements.

4.4.2 A Stroke-Based Approach for Simplifying Schematic Network Maps

Figure 4.23 shows an example of road network schematic maps: Tube Map London Underground. Such schematic maps are effective in conveying information in limited scope, making them easy to interpret by concentrating on the relevant aspects of the information and omitting unnecessary details based on linear abstractions of networks. It has been a dream of cartographers to produce such 'simple, clear not complex' maps by automated means.

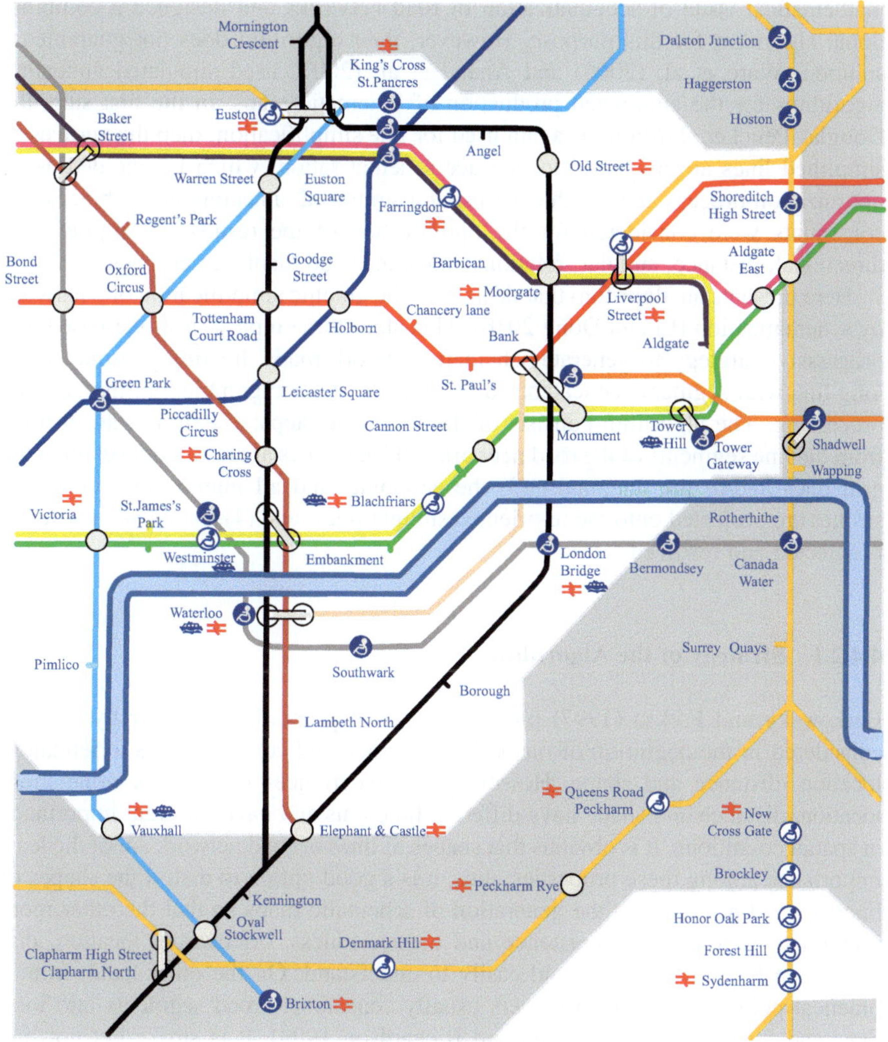

Fig. 4.23 An example of schematic road network maps: Tube Map London Underground

For the automated generation of schematic maps for underground routes (Avelar 2002, 2007), Scott and Rodgers (2005) employed a graph model. In this model, stations were represented by nodes and routes by edges. Elroi (1988) defined three steps for the schematization of general road networks, i.e. simplifying lines to basic shapes, re-orienting lines along grid lines, and enlarging those areas where roads are congested. Avelar (2007) has made use of the Douglas-Peucker Algorithm to remove the details of the lines and they adjusted the straightened linear network using a dynamic iterative algorithm. Cabello et al. (2001) have discussed the mathematical issue of schematization of road networks and designed a combinational algorithm for this purpose. However, their algorithm does not guarantee a solution. Ware et al. (2006) and Anand et al. (2007) used simulated annealing algorithms for the automated production of schematic maps. In the first step, the Douglas-Peucker algorithm is again used for line simplification; then the vertices of simplified lines are displaced to produce schematization, which is regarded as an optimization process and is achieved using a simulated annealing algorithm. Seven constraints were considered for the optimization of the results, i.e. topological, directional, distance, angular, rotational, clearance and displacement factors.

Here presents an algorithm that deals with the graphic simplification of a network by schematization (Li and Dong 2010). This algorithm employs a stroke-based and progressive strategy to generate schematic network maps, treating a stroke (i.e. a long line with segments concatenated together) as a basic graphic unit in map data processing. The algorithm consists of the following steps: (1) strokes are formed from the line segments of the road network; (2) the strokes are re-orientated along the grid lines and/or the diagonals; (3) the endpoints and all intersections on (sub-) strokes are projected onto the re-oriented straight lines; and (4) spatial inconsistency is detected and resolved.

4.4.2.1 Strategy of the Algorithm

Barkowsky and Freksa (1997) proposed a set of properties in hierarchy that are considered in the cognition of maps, i.e. existence and connectedness, orientation, location, distance and shape. However, our experiences tell us that orientation, location, distance and shape have different functions and have different importance in spatial cognitions. It is obvious that shapes of lines of road network are of the least importance among these properties; thus, it is a good option to distort the shapes of lines of road networks in the generation of schematic maps so that the other more important properties (i.e. existence and connectedness, orientation, location, distance) can be shown clearly and easily to understand. On the other hand, from a practical point of view, map readers usually concatenate road segments into long lines and form strokes (Thomson and Richardson 1999). It is stroke but not line segment that is used as the basic unit in schematization.

To guarantee that the result from every step satisfies the requirements described above, a 'progressive' strategy is adopted in the algorithm to generate schematic network maps by handling strokes one by one according to the hierarchical structure of the strokes.

4.4.2.2 Procedures of the Algorithm

The algorithm (Li and Dong 2010) includes the following four steps.

1. Formation of strokes from the line segments of the road network

When road segments are short they may be concatenated into long lines that are called "strokes". If the road segments have an identical name, they are usually linked and form a stroke. Otherwise, if the road segments have different names or do not have names, they can be concatenated to form strokes by several principles and rules, e.g. the principle of Good Continuation from the Gestalt psychology. Abiding by the principles and rules, strokes can be formulated step by step:

Firstly, set a deflection angle threshold. A value 45° is selected in this algorithm study, because it has been concluded that a deflection angle threshold between 30 and 75° can produce stable outcomes (Jiang et al. 2008), and this algorithm uses the four-direction system and the eight-direction system for reorientation of lines.

Secondly, for each line segment, compute all the deflection angles formed by itself and each of all the segments that intersect at the same point (node).

Thirdly, form a stroke with the given segments if none of these deflection angles is less than the threshold; or else add those segments with deflection angles less than the threshold into the set of so-called possible joining segments (PJS) for consideration.

Last, select a segment from the PJS based on one of the three principles of every-best-fit, self-best-fit and self-fit.

The self-fit is the laziest strategy as it randomly selects a segment from PJS as the next one (Fig. 4.24a). Here, line segments b, c and d but except e can be chosen as the next segment linking with line segment a if the deflection angle is less than a certain threshold. In self-best-fit, each segment is very selfish. It only considers the benefit to itself of choosing a segment from PJS where the deflection angle with itself is the least. In this case, a is connected to b (Fig. 4.24b). In every-best-fit, both the given segment and all the PJS are considered to find the best way to form strokes. In other words, every pair of segments at a junction point must negotiate with each other to have the best fit. In the case shown in Fig. 4.24c, although line segments a and b have the minimum deflection angle if viewed from line segment a, yet line segments b and e have a less deflection angle if viewed from segment b or e, and thus line segment a is connected to c and line segments b and e have the priority to connect each other.

2. Re-orientation of the strokes along the grid lines and/or the diagonals

Here, reorientation of the strokes refers to formatting long strokes in a given direction system. There are two direction systems in practice: the four-direction system and the eight-direction system. For the former, the horizontal and vertical directions are prior to other directions in the process of forming strokes; for the latter, horizontal, vertical and diagonals have equal priorities.

It is necessary to check whether a long stroke can be re-orientated to a single direction. If not, the stroke will be bent to fit into more than one direction. For example, Stroke 2 in Fig. 2.25b is fitted into three directions in the light of its running trend. Two approaches can be used to check such a necessity (see Fig. 4.25 for an example):

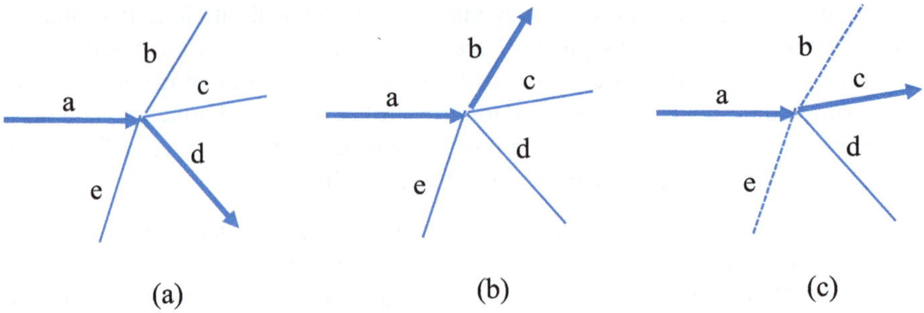

Fig. 4.24 Approach to selecting a segment for concatenation: (**a**) self-fit; (**b**) self-best-fit; and (**c**) every-best-fit

Approach 1 is for checking the maximum direction distortions: to check the possible maximum direction distortions against a given threshold (e.g. 45° if the four-direction system is employed), which are to be introduced if the straight line connecting the first and the last point of the stroke is used to represent the stroke itself. If no direction distortion is greater than the given threshold, then the stroke is straightened and oriented accordingly. Otherwise, two orientations should be considered.

Approach 2 is for checking the sequential direction changes: to check the direction changes sequentially. If a change in direction at a point is greater than the given threshold, then a change of orientation at this point is warranted.

3. The projection is to put the endpoints and intersections from the strokes onto the reoriented straight lines. After the re-orientation of a sub-stroke, the two endpoints of all intersection points and all intermediate points on this sub-stroke are then projected onto the straight line. As a result, original segment orientations of this stroke or sub-stroke will be re-orientated to the same direction; and then the two endpoints of all intersection points are projected and retained, while all intermediate points are removed to make the result simpler and clearer. Usually, two types of projection (Fig. 4.26) are employed, i.e. perpendicular projection and proportional projection.

The objective of the perpendicular projection is to project points perpendicularly onto the straight line. Figure 4.26a shows such a projection. In this way, original X-coordinates of all new points are not changed but their Y-coordinates changed into an identical value. It is also easy to design a formula to calculate the coordinates for the proportional projection (Fig. 4.26b).

Figure 4.27 shows an example in the four-direction system. If the road network is in the eight-direction system, they can usually be projected onto a horizontal line, a vertical line, a positive diagonal line, or a negative diagonal line. The first two cases are the same as that in the four-direction system, so only the last two ones are discussed in the following discussion.

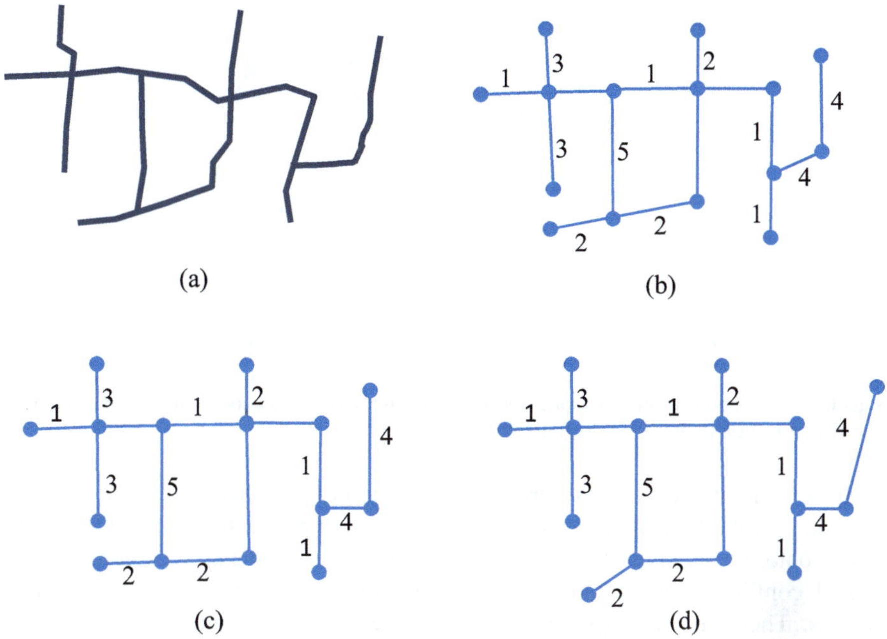

Fig. 4.25 Two options for re-orientation of strokes. (**a**) Original road network, (**b**) possible result 1: fitted and straightened, (**c**) possible result 2: in the four-direction system, and (**d**) possible result 3: in the eight-direction system

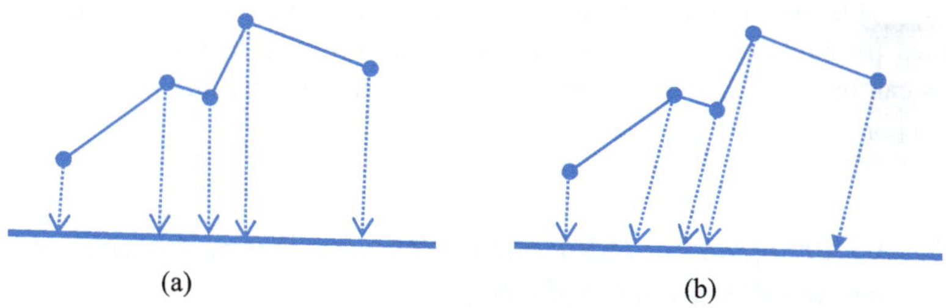

Fig. 4.26 Two types of projections for mapping points from sub-strokes onto re-oriented lines. (**a**) Perpendicular projection, and (**b**) proportional projection

In the case of projection onto a diagonal line, a three-step procedure can be applied to calculate coordinates:

Step 1: to rotate the sub-stroke by 45° (or −45°);

Step 2: to project the points from the sub-stroke onto the horizontal (or vertical) grid line; and

Step 3: to transform the projected coordinates back to the original coordinate system by inverse rotation.

4. Detection and resolve of spatial inconsistency

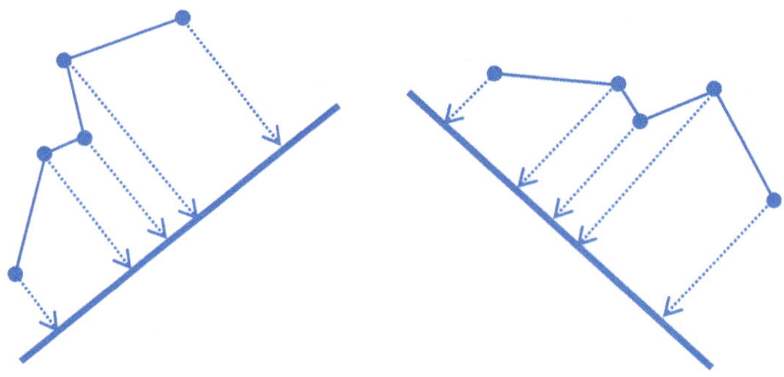

Fig. 4.27 Projections from sub-strokes onto diagonal lines. (**a**) Projection to the 45° diagonal line, and (**b**) projection to the −45° diagonal line

It is straightforward to re-orientate an individual (sub-)stroke onto a given direction (in the four-direction system or the eight-direction system) and to project points onto the direction. However, the re-orientation and the projection may cause spatial conflicts (mainly topological inconsistency) among/between neighboring strokes. Therefore, a progressive strategy is employed to guarantee topological consistency in the process of generating schematic network maps.

"Progressive" here means strokes are handled one by one in the schematization process, in the handling of the current stroke, the previously schematized strokes are treated as references. This guarantees that the results from each step for each stroke are reasonable. To achieve this, the strokes are put into an order according to their priorities (or rankings). In accordance with the order, a hierarchical map of the strokes can be formed, and the topological conflicts among roads can be avoid.

4.4.3 A Cartographic Information Theory-Based Algorithm for Road Network Selection

The selection of road segments from road networks is a fundamental and necessary operation in road network generalization, which is usually done before other generalization operations such as simplification, smoothing, caricature and displacement of roads. Indeed, it is a prerequisite of other generalization operations and is viewed as an inevitable and effective step in road network generalization. Because of the importance of this issue, many algorithms have been developed to solve the problem. They can be classified into three categories, i.e. semantic-based, graph-based, and stroke-based algorithms.

Semantic-based algorithms make decisions on selecting roads by their semantic attributes, such as street types and the number of lanes (Jiang and Claramunt 2004), and streets are generally selected in a ranked order according to their relative importance of attributes. However, past researchers' analyses have shown that semantic-based methods are not satisfactory, because geometric and topological information containing in road networks are usually not considered (Ai 2007). Graph-based algorithms view road networks as connected graphs. They usually work by incorporating some other mathematical tools such as shortest path, minimum-spanning-tree and degrees of centrality, and pay more attention on topological relationships among roads at mezzo spatial level (Mackaness and Beard 1993; Mackaness and Mackechnie 1999; Thomson and Richardson 1999; Ruas 2000; Jiang and Claramunt 2004; Jiang and Harrie 2004). Nevertheless, graph-based methods generally do not explicitly consider thematic and geometric information containing in road networks. Stroke-based algorithms construct strokes of the roads by the principle of "good continuation". Although they consider geometric, topological, and thematic aspects of the roads, they neglect statistical information of the roads and may generate poor results in some cases (e.g. road networks in sparse area).

To overcome the shortcomings of the three categories of algorithms, an algorithm (Liu et al. 2010) was proposed which considered the four types of information (i.e. statistical, metric, topological, and thematic information) at three scales (i.e. macro, mezzo and micro scales) in road network selection. At macro scale, the focus of information description is on general patterns of the road networks; at mezzo-scale, the focus is on relations among road segments; and at micro-scale the focus is on individual roads' properties. The algorithm mainly consists of three steps:

Step 1: measurement of the information containing in road networks,
Step 2: selection of roads,
Step 3: assessment of selection results.

4.4.3.1 Measurement of the Information Containing in Road Networks

To integrate the four types of information at the three different spatial levels in road network generalization, appropriate measures should be selected for quantifying each type of information so that every type of information can be described correctly. Table 4.1 lists such measures. Among these measures, the meanings of total number of roads, stroke length, number of intersections, stroke rank, deflection angel between roads, and distance between the end points of roads are obvious and thus it is unnecessary to further discuss them here. The following paragraph only discuss the concept "local road density" which is defined based on Voronoi diagrams.

Local road density and its general distribution are derived from a skeleton partitioning of the road network (Liu et al. 2009). The space among road segments is partitioned equally by skeleton with linked neighboring skeletons forming the Voronoi diagrams for road segments (Fig. 4.28). With the construction of these Voronoi polygons, the local road density of a road segment can be defined as the ratio between the length of the road segment and the area of the Voronoi polygon occupied by this road segment.

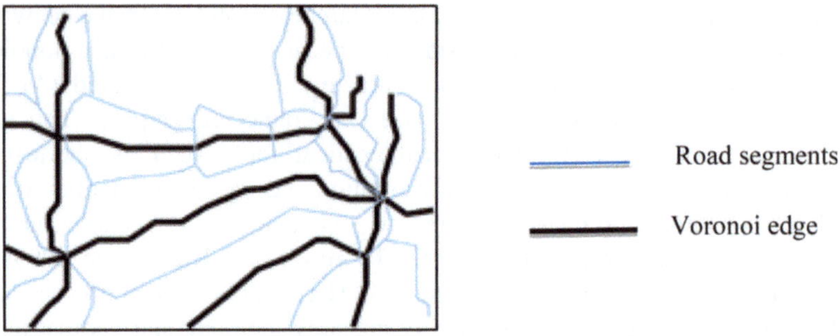

Fig. 4.28 Voronoi tessellations of a river basin

The Voronoi-based local road density is an indicator of local road density, because each bounding Voronoi polygon of the corresponding road can be deemed as the space that the road segment possesses. In this sense, roads which acquire smaller spaces are usually densely distributed, and thus more roads need to be eliminated during selection process.

After determination of the measures, strokes of the road network can be formed using a so-called "see extension" approach (Liu et al. 2009).

4.4.3.2 Selection of Roads

The road selection algorithm includes the following three major steps.

Firstly, the measures listed in Table 4.2 are calculated in light of the Voronoi diagrams and strokes of the road network, beside the Voronoi-based density, stroke-related measures, and other measures such as the number of road segments and the distribution of local road density before generalization.

Secondly, the road selection is executed taking each stroke as a selection unit and using the measures calculated in the first step. Strokes are clustered according to their weighted topological, thematic and metric properties. The resulting clusters are then ranked in descending order according to the cluster centers' values. Road segments contained in the first M (M = 1, 2, 3, . . .) clusters of strokes are selected when the total number of the strokes in the first M clusters is around the predefined number of segments to be selected.

Last, the selection result is assessed based on the local road density and its threshold. The road density threshold indicates the density distribution differences between the source and the generalized maps and can be derived from the comparison between the source map and the manually generalized map.

Table 4.2 Measures for describing the four types of information at three spatial levels

	Macro	Mezzo	Micro
Metric		Stroke length; number of intersections	Local density of roads
Thematic		Stroke rank	
Topological		Deflection angle between roads; distance between the endpoints of roads	
Statistical	Total number of roads; road density distribution		

4.4.3.3 Assessment of Selection Results

The results in the experiments (Liu et al. 2009) have shown that this algorithm is efficient and satisfactory.

The experiments have demonstrated that the strokes generated by the extension algorithm and that generated by cartographers are highly similar, with relatively small variance. This high similarity reveals that the seed extension algorithm is effective in generating strokes and the results are similar to those identified by human visual system.

The selection results in the experiments also have revealed that the general pattern and the density differences of the road network are maintained after selection. The differences between manual selection results and the results generated by the selection algorithm are not significant. Results also show that more roads are selected as well as deleted in the area with higher road density. In other words, more roads are deleted from lower-right corner of the region while most of the roads in the upper-left corner are selected. This is very interesting, and its reason might exist in human's spatial recognition.

4.4.4 Mesh Density-Based Approach for Selective Omission of Road Networks

The generalization process of road networks may require various operations such as selective omission, simplification, typification, collapse and displacement (Li 2006). Among these operations, selective omission usually precedes other operations (Jiang and Harrie 2004), which is a process to retain more important road features (or to eliminate less important ones) while the essential topological, geometric and semantic characteristics of the road network are preserved.

For the selective omission of roads, many approaches have been proposed. Mackaness and Beard (1993) used the graph theory to support generalization of road network. Mackaness (1995) and Jiang and Claramunt (2004) applied space syntax for the analysis of the structure of urban street networks (Porta et al. 2006).

Jiang and Harrie (2004) employed self-organizing maps to determine the significance of individual road features, which is a synthesized index of centrality measures and other geometric and semantic attributes of streets (such as length and type). Thomson and Richardson (1999) made use of the principle of "good continuation" from the Gestalt theory to concatenate road segments into strokes. Chen et al. (2009) proposed an algorithm for selective omission of road features by measuring the density of road network, i.e. road mesh density which is an important constraint in road generalization. This algorithm is called the mesh density-based approach here and will be discussed in detail in the following paragraphs.

The mesh density-based approach (Chen et al. 2009) calculate the density of a road network and its local variations using meshes as units. Since the features on maps at different scales usually have different densities, different density thresholds for road network generalization are determined by theoretical analyses and empirical studies of the mesh densities on the maps at different scales. The selective omission procedure is as follows: (1) It starts with the identification of the meshes that have a density beyond the threshold. (2) The mesh with the largest density is selected. Its bounding road segments are ordered according to their relative importance. Here, the least important segment is eliminated, and the remaining segments are then merged with the adjacent mesh, thus forming a new mesh. The selection procedure is repeated until none of the meshes has a density beyond the threshold.

4.4.4.1 Identification of Mesh Density and Its Threshold in Road Network Generalization

For road generalization based on the mesh density, a mesh can be regarded as the basic unit of selection. With a given set of roads, the mesh density can be calculated by Formulae 4.8, 4.9, and 4.10. The mesh density is then used as a constraint to determine which meshes should be selected. Usually, a threshold is given beforehand or computed from the given set of data. If the density of a mesh goes beyond the threshold, then there is a need to eliminate one or more road segments on its boundary. Thus, the determination of such a threshold is worth discussing in detail here.

Determination of the Threshold

The density threshold for each class of meshes can be determined by theoretical analysis or map specifications.

Firstly, as far as using theoretical analysis to get the threshold is concerned, mesh density can be expressed in terms of ground distance (see Formula 4.8). Therefore, it is a fixed value for a given mesh. If map distance but not ground distance is used, it may be calculated by:

$$D_m = \frac{L/S}{A/S^2} = \frac{L}{A}.S = D.S \tag{4.11}$$

Where, D_m is the density of the mesh; S is the denominator of the scale of the map; A is the area of the mesh; and L is the total length of the edges of the mesh.

Formula 4.11 reveals that the less the scale of the map, the greater the mesh density. On a certain scale map, there is a minimum size for a mesh unit below which the mesh cannot be discerned by human's eyes anymore. Correspondingly, there is a threshold of mesh density beyond which one or more segments on the boundary of the mesh must be eliminated so that two or more adjoining meshes could be merged to form a larger mesh. Such a threshold is regarded as the permissible largest density (represented by D_{m-max}) and may be calculated by

$$D_{m-max} = \left\{ \begin{array}{c} \pi L_m / \pi \left(\frac{L_m}{2}\right)^2 \; or \\ 4L_m / (L_m)^2 \end{array} \right\} = {}^4/_{L_m} \tag{4.12}$$

Where, L_m is the diameter of the smallest visible object in terms of map distance.

Secondly, as far as using map specifications to get the threshold is concerned, L_m is the diameter of the smallest areal symbols designated in map design and/or compilation specifications. Table 4.3 lists some examples of the smallest areal objects that usually used in China's map design specifications. Thus, it is easy to get the threshold value D_{m-max} by Formula 4.12.

Preservation of Density Variations Across Regions

The meshes can be classified based on the types of roads on its boundary. A mesh may be bordered by roads of different classes such as main streets (MS) and secondary streets (SS) (see Table 4.1). Then meshes are classified based on the bounding road segments. In this method, the class of a mesh is assigned with bounding segments of the lowest ranking, and different thresholds may be applied to different classes of meshes to preserve the density variations across different regions of a road network.

Table 4.3 the size of the smallest areal symbols on maps

Map symbol	Diameter of the symbol (threshold value)
■ Square	Edge: 0.4 mm
● Circle	Diameter: 0.4 mm
▲ Equal lateral triangle	Width: 0.4 mm
▲ Isosceles triangle	Width: 0.6 mm
▬ Rectangle	Width: 0.4 mm

Preservation of the Connectivity of a Road Network

When the density of a mesh exceeds the given threshold, one or more road segments on the mesh boundary should be eliminated. As a mesh is a loop, when one or more segments are omitted, the remainder becomes part of a new larger loop formed by one or more originally adjoining meshes. In practice, the aggregation is conducted successively by merging two adjoining meshes at a time. The relative importance of a certain road segment of a mesh depends on its relative contribution to the topological connectivity of the road network and its attributes such as road class, length and so on.

4.4.4.2 Selective Omission of Road Networks

Parameters Considered in the Approach

To determine which segments of the road network should be deleted, several parameters, such as the name, class, length of the road, need to be considered in the process of road map generalization. However, the values of these attributes are not always available. To solve this problem, the road length and the degree of strokes are used in this approach because they may be easily obtained, and more importantly they can reflect the geometric, semantic and topological characteristics of road segments.

In the process of manual selection, the visual length of a road plays a vital role: the longer a road in cartographers' eyes, the more probably it will be retained on the resulting map. Therefore "stroke" proposed by Thomson and Richardson (1999) is used here to achieve this idea. A stroke lays the foundation on the principle of "Good Continuation" in the theories of Gestalt psychology, which can group together the road segments that appear to follow in the same direction. The long linear elements formed by this principle are called strokes of roads. Once a stroke is obtained, its length and the "degree of the stroke" can be calculated. Here, the degree of a stroke is the number of segments in the stroke. The more degree of a stroke indicates a larger number of intersections the stroke has, and the more important the stroke is. For example, the degree '1' means that the stroke has the least relation with other segments in geometric and semantic sense, therefore, it is more probable for this stroke to be eliminated than other strokes.

Other parameters might also be useful. For example, Li and Choi (2002) prosed six thematic attributes of roads at different scales, i.e. type (or class), length, number of lanes, number of traffic directions, width and connectivity, to determine whether the road should be eliminated. In this approach, road length is substituted by stroke length, and the number of traffic directions is not always available and therefore is not used, and the connectivity is also considered through other means. To sum up, this approach considered road class, degree of stroke, length of stroke, length of segment, with the importance in descending order. In addition, sometimes, other implicit attributes are also considered. For example, a road segment linking directly

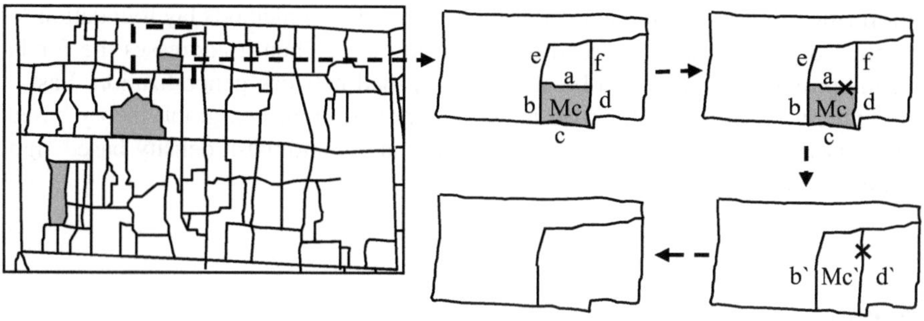

Fig. 4.29 Elimination of road segments and merging of adjoining meshes

with important features such as a dock should not be eliminated. The rules for elimination operation can be set based on those parameters.

Selective Elimination of Road Segments

The iterative process of eliminating road segments and merging adjoining meshes is illustrated in Fig. 4.29. In each iteration, the mesh with the largest density is identified and its least important road segment is eliminated. After completion of the elimination, the remaining road segments are re-organized, i.e. a mesh whose one or more edges have been eliminated should be merged with its topologically adjacent one to form a new mesh.

In the process of road segment deletion, several rules are employed.

Rule 1: The road segments of the lowest class should be eliminated first.

Rule 2: If the degree values of some strokes equal to 1, eliminate the shortest one of them.

Rule 3: If no degree of the strokes is equal to 1, search for the road segments to get the shortest one. If the shortest one is less than the threshold, eliminate it.

Rule 4: The road segments connecting with important human-made or natural features should be retained in the process of elimination.

Rule 5: If the length of a stroke is greater than a certain value, the road segments forming the stroke should not be deleted.

4.4.5 Summary of the Chapter

Road network is a type of important features. Multi-scale representation of road networks is necessary in constructing spatial data infrastructures. This chapter introduces the four categories of methods for describing road networks that can be found in literature, i.e. graph theory-based methods, stroke-based methods, information theory-based methods, and measures. These methods lay foundation for automated generalization of road networks on maps. Then it addresses the fundamental principles that should be abided by in road network generalization. Last, it presents five approaches for road network generalization, i.e., graph theory-based, stroke-based, cartographic information theory-based, and mesh density-based algorithms, respectively.

References

Agrawala M, and Stolte C., 2001, Rendering effective route maps: improving usability through generalization, in SIGGRAPH2001 Conference, Computer Graphics Proceedings, pp: 241–250, ACM Press.

Ai T., 2007, The drainage network extraction from contour lines for contour line generalization. ISPRS Journal of Photogrammetry and Remote Sensing 62, 93–103.

Anand S, Avelar S, Ware J. M and Jackson M, 2007, Automated schematic map production using Simulated Annealing and Gradient Descent approaches, 15th Annual GIS Research UK Conference (GISRUK 2007), Dublin, Ireland.

Avelar S. 2002. Schematic Maps on Demand: Design, Modeling and Visualization. Dissertation for Doctor's Degree. Zurich: Swiss Federal Institute of Technology, Switzerland.

Avelar, S., 2007. Convergence analysis and quality criteria for an iterative schematization of networks. Geoinformatica, 11(4): 497–513

Barkowsky T. and Freksa C., 1997, Cognitive Requirements on Making and Interpreting Maps, In: Hirtle S.C., Frank A.U. (eds) Spatial Information Theory A Theoretical Basis for GIS. COSIT 1997. Lecture Notes in Computer Science, vol 1329. Springer, Berlin, Heidelberg.

Benz S. A. and Weibel R., 2013, Road network selection using an extended stroke-mesh combination algorithm, paper orally presented at the workshop of the 16th ICA Generalisation Workshop, Dresden, pp: 23–24.

Cabello J.S., de Berg M.T., van Dijk S.F., van Kreveld M.J. and Strijk T.W., 2001. Schematization of road networks. In: Diane L. Souvaine Ed., Proceedings of the 17th annual symposium on computational geometry, 3–5 June 2001. Medford, MA: ACM, 33–39.

Chaudhry O. and Mackaness W.A., 2005, Rural and urban road network generalisation deriving 1:250, 000 from OS mastermap, in the Proceedings of the 22th International Cartographic Conference, La Coruña, July 10–16, 2005.

Chen J., Hu Y., Li Z., Zhao R., & Meng L., 2009, Selective omission of road features based on mesh, density for automatic map generalization, International Journal of Geographical Information Science, 23(8): 1013–1032

De Serres B. and Roy A.G., 1990, Flow direction and branching geometry at junctions in Dendritic River Networks, The Professional Geographer, 42(2): 149–201

Edwardes A.J. and Mackaness W.A., 2000, Intelligent generalisation of urban road networks, in the Proceedings of GIS Research UK 2000 Conference, York, United Kingdom, April 5–7, 2018.

Elroi D.S., 1988, Designing a network line map schematization software enhancement package, In the Proceedings of the 8th Anniversary of ESRI User Conference, Redlands, California, USA.

Ester M., Kriegel H.P. Sander J. and Xu X., 1996, A density-based algorithm for discovering clusters in large spatial databases with noise, in the Proceedings of the 2nd International Conference on Knowledge Discovery and Data Mining (KDD-96), pp:226–231, Portland, Oregon, USA, August 2–4, 1996. AAAI Press.

Haggett P., 1967, Network models in geography, in: ChorJey R.J. and Haggett P. (eds), Models in Geography, pp: 609–668, London: Methuen, UK.

Harrie L. and Weibel R., 2007, Modelling the overall process of generalization, In Generalisation of Geographic Information: Cartographic Modelling and Applications, edited by W. A. Mackaness, A. Ruas, and L. T. Sarjakoski, pp:67–88. Oxford: Elsevier.

Hart P.E. Nilsson N.J. and Raphael B., 1968, A formal basis for the heuristic determination of minimum cost paths, IEEE Transactions on Systems Science and Cybernetics 4 (2): 100–107.

Heinzle F. and Anders K.H., 2007, Characterising space via pattern recognition techniques: identifying patterns information: cartographic modelling and applications, edited by W. A. Mackaness, A. Ruas, and L. T. Sarjakoski, pp: 233–253. Oxford: Elsevier.

Heinzle F. Anders K.H. and Sester M., 2005, Graph based approaches for recognition of patterns and implicit information in road networks, in the Proceedings of the 22th International Cartographic Conference, La Coruña, July 10–16, 2005.

Jiang B. and Harrie L., 2004, Selection of streets from a network using self-organizing maps. Transactions in GIS, 8(3): 335–350.

Jiang B. and Claramunt C., 2004, A structural approach to the model generalization of an urban street network. GeoInformatica, 8(2): 157–171.

Jiang B. 2004, A Structure Approach to the Model Generalization of An Urban Street Network. Geoinformatica,(8):38–54

Jiang, B. 2009. "Street Hierarchies: A Minority of Streets Account for a Majority of Traffic Flow." International Journal of Geographical Information Science 23 (8): 1033–1048. doi:https://doi.org/10.1080/13658810802004648.

Jiang, B., Zhao, S., and Yin, J., 2008. Self-organized natural roads for predicting traffic flow: a sensitivity study. Journal of Statistical Mechanics: Theory and Experiment. DOI: https://doi.org/10.1088/1742-5468/2008/07/P07008.

Li Z., 2006, Algorithmic foundation of multi-scale spatial representation, New York: CRC Press.

Li Z.L. and Choi Y.H., 2002, Topographic map generalization: association of road elimination with thematic attributes. The Cartographical Journal, 39(2): 153–166.

Li Z. and Dong W., 2010, A stroke-based method for automated generation of schematic network maps, International Journal of Geographical Information Science, 24(11): 1631–1647.

Li, Z., and Q. Zhou. 2012. "Integration of Linear and Areal Hierarchies for Continuous Multi-Scale Representation of Road Networks." International Journal of Geographical Information Science 26 (5): 855–880. doi:https://doi.org/10.1080/13658816.2011.616861.

Liu X., Zhan F., Ai T., 2010, Road selection based on Voronoi diagrams and "strokes" in map generalization, International Journal of Applied Earth Observation and Geoinformation, 12(4): 194–S202

Liu, X., Ai, T., Liu, Y., 2009. Road density analysis based on skeleton partitioning for road generalization. Geo-Spatial Information Science 12, 110–116.

Mackaness W.A. and Beard, M.K., 1993, Use of graph theory to support map generalization. Cartography and Geographic Information Systems, 20(4): 210–221.

Mackaness W.A., 1995, Analysis of urban road networks to support cartographic generalization. Cartography and Geographic Information Systems, 22(4): 306–316.

Mackaness W.A. and Mackechnie G.A., 1999, Automating the detection and simplification of junctions in road networks, Geoinformatica, 3 (2): 185–200.

Morisset B. and Ruas A., 1997, Simulation and agent modelling for road selection in generalization, in the Proceedings of the 18th International Cartographic Conference, Stockholm, Sweden, June 23–27, 1997.

Newman M.E.J., 2003, The structure and function of complex networks. SIAM Review, 45, 167–256.

Porta S. Crucitti P. and Latora V., 2006, The network analysis of urban streets: a dual approach, Physica A: Statistical Mechanics and Its Applications, 369 (2): 853–866.

Ren C. Fang H. Lei Y. Tian J. and Yang W., 2015, Network functionality oriented stroke building in road networks. In the Proceedings of 23rd International Conference on Geoinformatics, Wuhan, China.

Richardson D.E. and R.C. Thomson, 1996, Integrating thematic, geometric, and topological information in the generalization of road networks, Cartographica, 33 (1): 75–83.

Ruas A., 2000. The role of mezzo object for generalization. In: Proceedings of the 9th International Symposium on Spatial Data Handling, Beijing, China, pp. 50–63.

Scott J.M. and Rodgers P., 2005, Automatic metro map design techniques. In the Proceedings of the 17th Annual Symposium on Computational Geometry, 3–5 June 2001. Medford, MA: ACM

Thom S., 2005, A Strategy for collapsing OS integrated transport network (Tm) dual carriageways, Paper orally presented at the ICA Workshop on Generalisation and Multiple Representation, La Coruña, July 7–8, 2005.

Thomson R C. 2006, The "stroke" concept in geographic network generalization and analysis, in the Proceedings of 12th International Symposium on Spatial Data Handling, Vienna, Austria, pp: 681–697.

Thomson R.C. and Richardson D.E. 1999, The 'good continuation' principle of perceptual organization applied to the generalization of road networks, in the proceedings of the 19th International Cartographic Conference, Ottawa, August 14–21, 1999.

Thomson R.C. and R. Brooks, 2000, Efficient generalisation and abstraction of network data using perceptual grouping, in the Proceedings of the 5th International Conference on GeoComputation, pp: 23–25 (Accessed November 26, 2017 via http://www.geocomputation.org/2000/GC029/Gc029.htm)

Tomko M. Winter S. and Claramunt C., 2008, Experiential hierarchies of streets, Computers, Environment and Urban Systems, 32 (1): 41–52.

Touya G., 2010, A road network selection process based on data enrichment and structure detection." Transactions in GIS, 14 (5): 595–614.

Walter V., 2008, Automatic interpretation of vector databases with a raster-based algorithm, in the Proceedings of he International Archives of the Photogrammetry, Remote Sensing and Spatial Information Sciences, Vol. XXXVII, Part B2, Commission 2, Beijing, July 3–11. (accessible on December 26, 2017 via http://www.isprs.org/proceedings/XXXVII/congress/tc2.aspx).

Ware J.M. Anand S. Taylor G.E. and Thomas N., 2006, Automated production of schematic maps for mobile applications, Transactions in GIS, 10(1): 25–42.

Weibel R. and Dutton G.H., 1998, Constraint-based automated map generalization, in the Proceedings of the 8th International Symposium on Spatial Data Handling, Vancouver, Canada, July 11–15, 1998.

Weiss R. and Weibel R., 2013, Road network selection for small-scale maps using an improved centrality approach, Paper orally presented at the workshop on the 16th International Cartographic Conference: the Generalisation Workshop, Dresden, Germany, August 23–24, 2013.

Yan H. and Weibel R., 2008, An algorithm for point cluster generalization based on the Voronoi diagram. Computers & GeoSciences, 34(8): 939–954.

Yu X., 2001, Road Network simplification with knowledge-based spatial analysis, Supplement Journal of Geographical Sciences, 11:54–62.

Yang B. Luan X and Li Q., 2011, Generating hierarchical strokes from urban street networks Bbsed on spatial pattern recognition, International Journal of Geographical Information Science, 25 (12): 2025–2050.

Zhang H. and Li Z., 2011, Weighted ego network for forming hierarchical structure of road networks, International Journal of Geographical Information Science, 25(2), 255–272.

Zhang Q., 2004, Modeling structure and patterns in road network generalization, Paper presented at the workshop on Generalisation and Multiple Representation, Leicester, UK, August 20–21, 2004.

Zhou Q. and Li Z., 2012, A comparative study of various strategies to concatenate road segments into strokes for map generalization, International Journal of Geographical Information Science, 26(4): 691–715.

Chapter 5
Description and Generalization of River Networks

5.1 Introduction to River Networks

What does a tree and a river have in common in structure? The answer is rather obvious: a river's skeleton on the map is a tree-like structure. But why rivers are tree-like in structure?

Suppose we need to send supply of water from many places (the black dots in Fig. 5.1) lying within an extended area to a well (the red dot at the lower left corner). Different strategies may be devised to achieve this goal. The following example manifests that a tree-like structure is an efficient structure, the most efficient in fact, to convey matter (or energy) from an extended source to a single outlet.

Strategy 1 (Fig. 5.1a): each point sends its share to a selected nearest-neighbor according to a given pattern (e.g. Hamiltonian walk). This is a globally efficient system, as the shipment is carried out according to an ordered path which does not allow for superposition that would lower the efficiency of the system. On the other hand, it is also locally inefficient as, for instance, water from point immediately at the north of the destination could flow along a direction much shorter southward path without making a long detour.

Strategy 2 (Fig. 5.1b): each point individually chooses the shortest route, which makes the system locally efficient, but globally highly costly as the total distance traveled by the entire system is much greater than that in the strategy 1, due to the many superpositions (dashed as well as solid lines have been used to enhance clarity).

Strategy 3 (Fig. 5.1c): A reasonable trade-off between the two opposite strategies is provided by the tree-like structure shown in Fig. 5.1c where each point attempts to find the shortest path but repetition is avoided through a hierarchical construction, so that a global efficiency can be achieved. A quantitative mathematical proof of the above statement can be obtained by showing that a tree-like structure is the one minimizing both the average individual path and the total traveled distance. In this respect, a river network is obviously a two-dimensional projection of the three-

H. Yan, *Description Approaches and Automated Generalization Algorithms for Groups of Map Objects*, https://doi.org/10.1007/978-981-13-3678-2_5

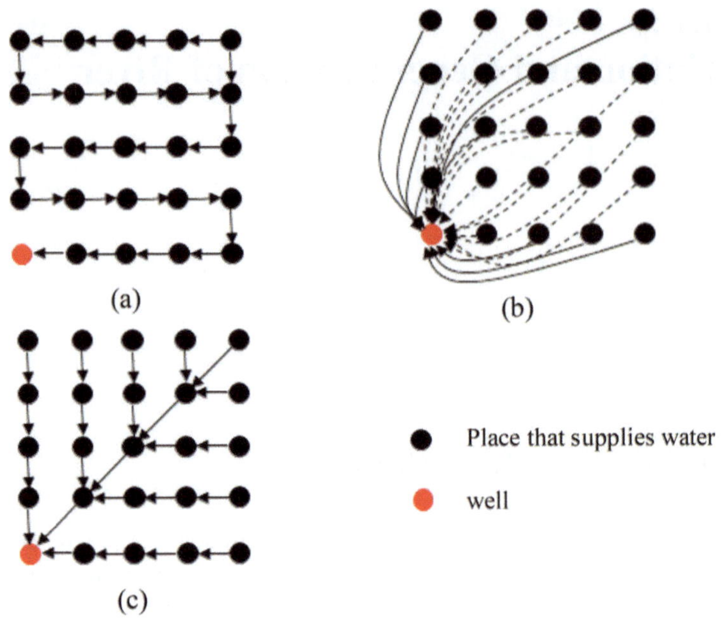

Fig. 5.1 Three structures with different efficiency in transferring water

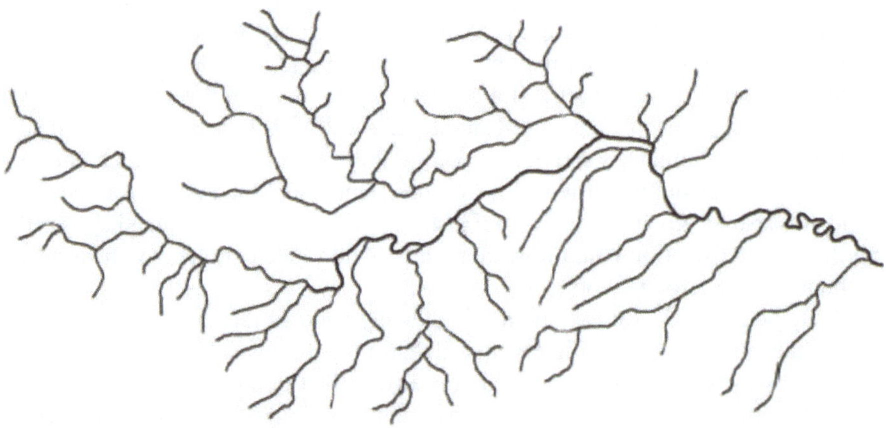

Fig. 5.2 River system: irregular networks on maps

dimensional tree-like landscape morphology, not necessarily (although frequently) related with the actual river flowing on that landscape in real geographic space, which is composed of the collection of all the paths formed by every tributary of the main river in its drainage basin.

Rivers are usually irregular networks on maps (Fig. 5.2) and are viewed as the skeleton of the terrain (Chen et al. 2012) and play important roles on multiscale representations on maps (Li et al. 2010; Génevaux et al. 2013). Recently, Attentions have payed to geographical features of river networks in the process of map generalization (Ai et al. 2006; Stanislawski 2009; Buttenfield et al. 2010) which follows the idea "generalization is not a mere reduction of information–the challenge is one of preserving the geographic meaning" (Bard and Ruas 2005). Therefore, this chapter emphasizes on discussing the approaches for describing and generalizing river networks, but ignores the factors, such as topography, soil and bedrock influencing the patterns of rivers networks that have been studied in detail for decades (Twidale 2004; Charlton 2007; Ritter 2016).

5.2 Descriptions of River Networks

Water bodies on the Earth include natural and manmade objects such as rivers, lakes, seas, wells, fountains, reservoirs, ponds, ditches etc. Because rivers are the terrain lines or character lines that "control" the other features on maps, only rivers are selected and how to represents and describes river networks will be addressed here.

A river basin consists of its mainstream and tributaries. Rivers on maps can be classified into four categories in the light of their geometric patterns, i.e. tree-like, grid-like, feather-like, parallel, radial, fan-like, comb-like, net-like, ring-like and centripetal (Fig. 5.3).

Tree-like river basin: the mainstream and its tributaries compose a tree-like graph. This is the most general kinds of rivers, for example, River Guijiang and River Yujiang in Guangxi Province, China.

Grid-like river basin: the mainstream and the tributaries perpendicularly intersect with each other and form the grid-like graph. A typical example of grid-like river basin is the Minjiang River in China.

Feather-like river basin: the tributaries are evenly distributed at the two sides of the mainstream, and the graph of the river on the map looks like a feather.

Parallel river basin: the tributaries are approximately parallel and afflux into the mainstream.

Fan-like river basin: the outline formed by the mainstream and the tributaries is like a fan, for example Haihe River, Yongdinghe River.

Comb-like river basin: most of the tributaries are at one side of the mainstream but few or none tributaries are at the other side which makes the graph of the river basin like a comb.

Radial river basin: such river basins usually appear around the volcanoes.

Centripetal river basin: they appear in depressional areas.

In literature, river basins on maps can be described using measures and by special methods for specific purposes which will be addressed in the following sections.

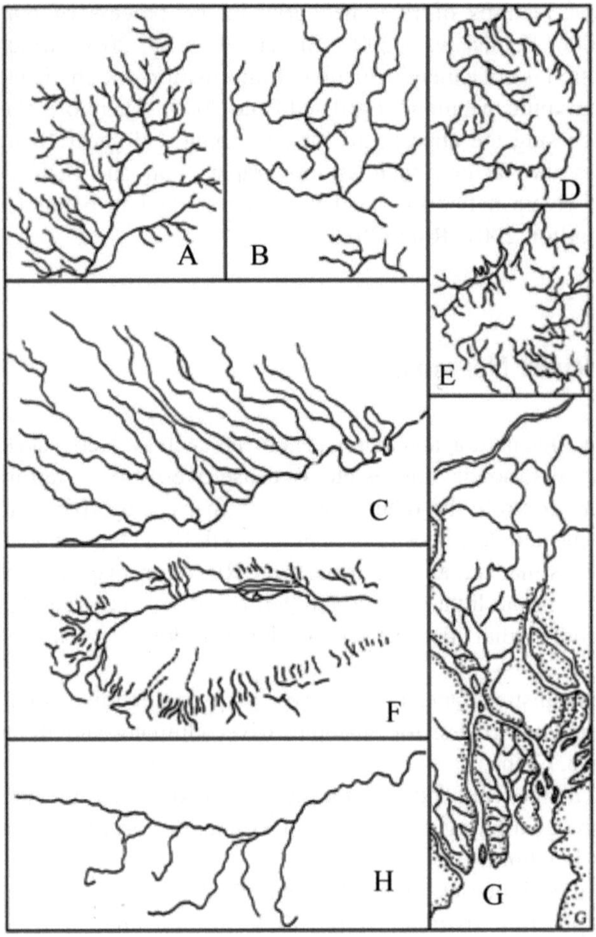

Fig. 5.3 Various patterns of river networks on maps

5.2.1 Measures and Parameters

Many measures and parameters have been developed for geometrically describing rivers.

- Density of the river network

  It can be calculated by Formula 5.1.

$$\mathrm{D} = \frac{\sum L}{F} \tag{5.1}$$

Where, D is the density of the river network; $\sum L$ is the total length of the tributaries and the mainstream of the river network; and F is the area of the river basin.

- Development coefficient of the river basin

It is a ratio between the total length of the tributaries and that of the mainstream.

$$M_{coe}^{i} = \frac{\sum l_i}{L_i} \tag{5.2}$$

Where, M_{coe}^{i} is the development coefficient of the ith order tributaries of the river network; L_i is the total length of the ith order mainstreams of the river network; and l_i is the total length of the ith order tributaries of the river network.

The greater M_{coe}^{i}, the greater the total length than that of the mainstream, the more developed the river is.

- Nonuniform coefficient of the river basin

It is the ratio of the total length of the tributaries on the left side of the mainstream and that on the right side of the mainstream. It can be calculated by Formula 5.3.

$$B_{coe} = \frac{\sum L}{\sum R} \tag{5.3}$$

Where, B_{coe} is the nonuniform coefficient of the river basin; $\sum L$ is the total length of the tributaries on the left side of the mainstream; and $\sum R$ is the total length of the tributaries on the left side of the mainstream.

The greater the difference between B_{coe} and 1, the more unbalanced the tributaries at the two sides of the mainstream.

- Ratio of the lakes/marshes

It is the ratio of the area of lakes/marshes and the total area of the river basin.

$$M_{ratio} = \frac{\sum M}{F} \tag{5.4}$$

Where, M_{ratio} is the ratio of the lakes/marshes in the river basin; $\sum M$ is the total area of the lakes/marshes; and F is the total area of the river basin.

- Length of the river segment

Rivers on digital maps are usually expressed using coordinates, and therefore the length of each river segment can be calculated by Formula (5.5).

$$d = \sum_{i=1}^{n} \left[(x_{i+1} - x_i)^2 + (y_{i+1} - y_i)^2 \right]^{1/2} \quad (5.5)$$

where, d is the length of the river segment; and (x_i, y_i) is the coordinates of the ith point on the river segment.

- Average length of the river segments of the river basin

It is a statistical value. If a river basin consists of m river segments, its average length of the river segments can be calculated by Formula 5.6.

$$\bar{d} = \left(\sum_{k=1}^{m} d_k \right) / \mathrm{m} \quad (5.6)$$

Where, $\bar{d}$ is the average length of the river segments of the river basin; and $d_k$is the length of the ith river segment.

- Angle between two river segments

About 88% of the angles between river segments are acute angles (De Serres and Roy 1990), which statistically tells the characters of river segments at intersections. Such angle values can be used in river classification, for the angles of different classes of rivers are different, e.g. the angles of tree-like rivers are acute and that of the grid-like rivers are approximately right angles.

- Meandering coefficient

It is a value used for describing the curvature degree of the linear river. To be simple, the meandering coefficient of a river can be the ratio between the total length the river and the distance of the straight line from the start point to the endpoint of the river.

- Flow direction of the river

The flow direction of the river indicates the terrain of the river basin. The flow direction of the mainstream is the predominant direction which indicates the tendency of the river network. Different categories of rivers can be differentiated according to their flow directions, for example, the flow directions of the mainstreams of a parallel river system are similar, and that of their first order tributaries are also similar.

- The number of the classes of the river basin

It means the number of the classes of the tributaries and the mainstream of the river basin.

- Interval of the river segments

It is the distance between the two adjacent river segments. This parameter indicates the density of river segments in an area. This value is usually obtained by measuring the distances of the river segments at the same side of a mainstream. It may be either the distance between two approximately parallel river segments, or the distance between two non-parallel river segments. Apparently, the latter is more common in the geographic space and more difficult to calculate.

- Watershed and catchment area

  The watershed of the river basin is the area that the tributaries flow through. The catchment area is the area of the watershed.

5.2.2 *Methods for Describing River Networks*

A measure or a parameter only gives a simple description of river networks, indicating a single aspect of the river, while a method that will be addressed below usually can provide systematic and comprehensive descriptions of rivers.

5.2.2.1 Information Entropy of River Networks

Informational entropy of river networks that is consistent with the definition given by Shannon (1948) in the information theory has been defined by Fiorentino and Claps (1992) in Formula 5.7.

$$\mathrm{s} = -\sum_{\delta=1}^{\Delta} P_\delta \ln P_\delta \tag{5.7}$$

where, s is the information entropy; and $P_\delta$is the relative probabilities related to a given state δ.In the definition of informational entropy of a river network, the network is considered as a system in which stream segments (links) are the elements whose placement characterize the system configuration. A network link is the path connecting two junctions and the topological distance from the outlet, i.e. the number of consecutive links forming the shortest path from its upstream node to the outlet, corresponding to the state δ in which the link is placed (see Fig. 5.4). The total number of states is the network topological diameter Δ, corresponding to the maximum topological distance from the outlet. If one disregards the length of links, the network configuration is completely determined by the topological width function which is the diagram of the relative frequency $P_\delta$of the links as a function of the topological distance δ. The maximum entropy for a given topological diameter is attained with uniform width function (i.e. same number of links at each topological distance) and is expressed as $S_{max} = \ln \Delta$.

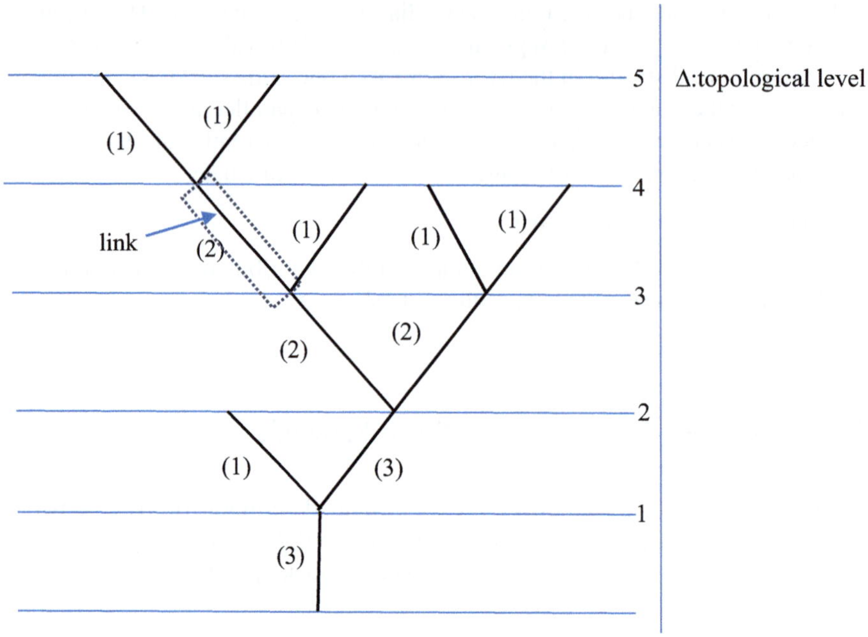

Fig. 5.4 Topological levels of river networks

Fiorentino et al. (1993) found that the informational entropy of river network varies with the regularity of network characteristics, particularly Horton order, magnitude and topological diameter. They expressed the entropy using fractal plan trees. Fractal plane trees are fractal (self-similar) objects, that can be defined as "having a shape made of parts similar to the whole in some way" (Mandelbrot 1986). Fractal objects result from repeated generations, starting from an initiator (i.e. a set of segments) and using a generator (i.e. a different set of segments). In the construction of the fractal set, the initiator is firstly substituted with the generator, and then each segment of the generator becomes an initiator and is substituted again in a recursive way (e.g. Feder 1988, p.16).

5.2.2.2 Fractal Dimensions of River Networks

Ever since Mandelbrot (1983) coined the term "fractal", there has been speculation that river networks are fractals. Fractals provide a mathematical framework for treatment of irregular, seemingly complex shapes that display similar patterns over a range of scales. Many objects in nature possess a property called statistical self-similarity. This may be defined as invariance of the probability distributions describing the object's composition under simple geometric transformational change of scale. Tarboton et al. (1998) deem that river networks fall into this class of geometric objects and that the fractal dimension characterizing the self-similarity of river networks is close to 2.

Hydrologists and geomorphologists have speculated on the fractal nature of river networks in the light of their indirect empirical evidence. One of the typical examples is Gray's method (1961) that tells the relationships between mainstream length (L) and basin area (A), presented in Formula 5.8.

$$\mathrm{L} = 1.4A^{0.568} \tag{5.8}$$

Mandelbrot (1983) speculated that river networks are fractal and got the fractal dimension D = 1.2 using Formula 5.8, taking the exponent as 0.6 but not 0.568. If individual rivers rather than river networks are concerned, $\mathrm{D} = 0.568 \times 2 \approx 1.1$.

Mandelbrot (1983) also studied some fractal geometric patterns that resemble river networks and got that the fractal dimension of these individual lines is 1.1, too. However, if the complete network patterns which can be viewed as the models of river networks are studied, he concluded D = 2 by the following reasoning.

If a line's shape (e.g. a coastal line or a stream) is considered, measure the length of the line (say, L) using a divider or a ruler whose length is r.

$$\mathrm{L} = \lim_{r \to 0} N.r \text{ or } \mathrm{N} \approx \mathrm{L}.r^{-1} \tag{5.9}$$

Where, N is the number of divider steps.

It is apparent that the less r which means the shorter the ruler, the more accurate L and N. However, Formula 5.9 are not convergent, because the implied exponent of r is 1.

If the implied exponent can be a fraction, a measure F which is independent of r can be obtained,

$$\mathrm{F} = \mathrm{N}.r^{D} = constant \tag{5.10}$$

Where, D > 1 is called the fractal dimension by Mandelbrot (1983).

This implies:

$$\mathrm{N}\tilde{\ } r^{-D} \tag{5.11}$$

or equivalently

$$\mathrm{L}\tilde{\ } r^{1-D} \tag{5.12}$$

Based on Mandelbrot's work, Lovejoy et al. (1987) proposed another technique to estimate the fractal dimension of river networks by functional box counting. It works on a set of points embedded in a d = 2 dimensionalspace. Cover the space with a mesh of d-dimensional cubes of size r^d. A relationship of the number (N(r)) of cubes that contain elements of the set is as follows:

$$\mathrm{N(r)}\tilde{\ } r^{-D} \tag{5.13}$$

This formula is based on a definition of fractal dimension given by Hentschel and Procaccia (1983) as

$$\mathrm{D} = -\lim_{r \to 0} \lim_{m \to \infty} \frac{\log N(r)}{\log r} \tag{5.14}$$

Where, m is the number of points in the set. The evidence given by Tarboton et al. (1998) strongly supports that the river network is space filling with D = 2.

The above fractal theory is closely related to Horton (1945) laws of network composition.

In Horton's laws, the length and bifurcation ratios are usually stated in terms of Strahler's (1952) ordering scheme. Source streams are at the first order. If two first-order streams join, they become second order; and generally, if two streams of equal order (say, the order equals k) merge, a stream at order k+1 is formed. When a lower and a higher order streams join, the successive stream retains the order of the higher-order stream. The set of empirical laws collectively referred to as Horton's laws include two formulae regarding to bifurcations (i.e. Formula 5.15) and lengths (i.e. Formula 5.16) of river networks:

$$R_b = \frac{N_{\omega-1}}{N_\omega} \tag{5.15}$$

$$R_l = \frac{L_\omega}{L_{\omega-1}} \tag{5.16}$$

Where, N_ω is the number of the streams of order ω. L_ω is the mean length of the streams of order ω.

Formulae 5.15 and 5.16 reveal the geometric-scaling relationships in river networks, since they hold no matter at what order or resolution human beings view the networks. If people regard a channel network as the paths where waterflows, people may imagine, if the resolution becomes higher and higher and the orders of the streams become lower and lower, they are literally looking at flows among the grass roots. Viewed this way, the limiting channel network is a fractal with properties governed by R_b and R_l.

Starting from Horton's laws, LaBarbera and Rosso (1987) found the fractal dimension of river networks may be calculated by

$$\mathrm{D} = \max\left(\frac{\log R_b}{\log R_l}, 1\right) \tag{5.17}$$

This formula requires that Horton's bifurcation and length ratios hold exactly at all scales in the network. The result is obtained by considering the limit of this series which converges for $R_b < R_l$implying D = 1. However, if $R_b > R_l$, the series diverges, and the total length of channel networks follows

$$L\sim S^{1-(\log R_b/\log R_l)} \tag{5.18}$$

Where, the resolution of observation of the networks is taken as the length of first-order streams, and

$$D = \frac{\log R_b}{\log R_l} \tag{5.19}$$

To sum up, river networks can be viewed as fractal, and existing techniques in estimating fractal dimensions all tend to indicate that the fractal dimension of river networks is 2. This is consistent with the fact that river drain the entire catchment basin and thus space filling, which provides a fundamental link between Horton's ratios. It is also worth noting that the random topology model proposed by Shreve (1967) is support D = 2, because it gave average values $R_b = 4$ and $R_l = 2$. The view of river networks as fractal with D = 2 therefore provides a description of the scaling of river networks that is consistent with classical fluvial geomorphology and the popular random topology model.

5.2.2.3 The River Tree

It is common to view rivers on maps as trees because of their dendritic structures; thus, many scholars describe rivers using so-called tree structure from computer sciences (Zhang and Quan 2005; Zhang 2006). The following presents an algorithm for constructing river trees, which is an improved version of the method proposed by Zhang and Quan (2005).

Step 1: organize the river data to form a direction graph

The river segments are viewed as the edges of the graph and the starting and ending points of the river edges are viewed as vertices. By this way, the river can naturally be regarded as a directed graph. Each edge of the graph contains two directed edges, with one's direction being the same as that of the river and the other's direction being counter that of the river.

The river is divided into river segments and re-organized to form a directed graph. The relations among the vertices and the edges are also recorded (e.g. topologically adjacent relations among edges, joint relations among edges and their starting and ending points). Meanwhile, the estuary (i.e. river mouth) of the river is marked and saved.

Step 2: determine and re-organize the directions of the river segments

The river data are usually collected starting from upstream to downstream; however, it is not always the case. Thus, it is necessary to judge the directions of the river segments. If the direction of a river segment is saved inconsistently to that of the river, its vertices should be reversed.

Step 3: Construct the river entities

The purpose of this step is to link the river segments to form a river entity.

Two rules that should be abided by in this step are: (1) the river segments with the same river name should be merged into the same river entity; and (2) two river segments with different flow directions may not be linked.

Nevertheless, not all river data have river names in practice. Hence, if the river segments have names, they are linked to form entities by their river names; otherwise, their geometric characters should be considered. Because the longer a branch is, the more probably it might be viewed as the mainstream of the river, calculations of the lengths of the river segments need to be done firstly. Then, (1) search and find the origin of the river from the river segments and clearly mark the starting point and the original river segment (it is named 'current river segment') of the river entity. (2) Start from the current river segment to find the next river segment that can be linked. Check the degree of the ending vertex of the current river segment. If the degree is 1, the river entity is completed and this procedure can be ended. If the degree is 2, the current river segment has only one river segment at its downstream. So, link them and take this river segment at the downstream as a new 'current river segment'. If the degree is greater than 2, the current river segment has one river segment at its downstream and more than one river segments at its upstream. In this case, if the river segments have names, they can be linked according to their names (i.e. the river segments with the same river names can be merged); otherwise, link the current river segment with the longest one of the river segments at upstream and take the longest river segment as the new 'current river segment' (see Fig. 5.5). (3) recursively execute the procedure in (2) until all river segments are marked.

Step 4: Differentiate between right and left branches of the river tree

Suppose that the mainstream and the two tributaries are saved in a spatial database, including their vertices and flow directions, the following method can be used to tell apart the right and the left tributaries by comparing the three angles α,β and γ, and any three angles in $[0^\circ, 360^\circ]$. If one of the two tributaries are considered, we have

α:the angle starting from the positive direction of the x axis and counter clock-wisely rotating to the opposite direction of the tributary.

β:the angle starting from the positive direction of the x axis and counter clock-wisely rotating to the opposite direction of the upstream of the mainstream.

γ:the angle starting from the positive direction of the x axis and counter clock-wisely rotating to the downstream of the mainstream.

If $(\beta - \gamma) > (\alpha - \gamma)$the tributary is at the right of the mainstream; otherwise, the tributary is at the left of the mainstream.

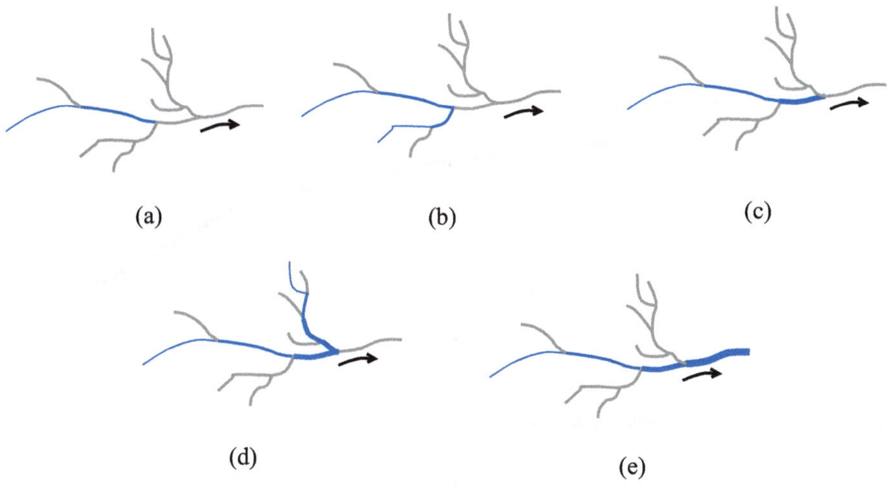

Fig. 5.5 Construction of the river entity from upstream to downstream

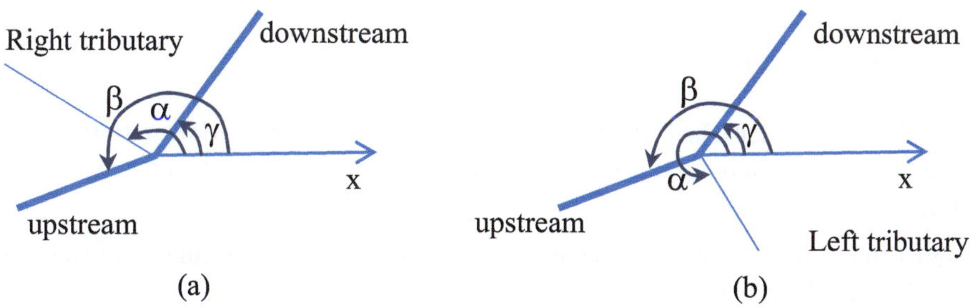

Fig. 5.6 Four cases for demonstrating how to determine the left and right tributaries of the mainstream

Figure 5.6 presents the method for telling apart left and right tributaries of mainstreams.

5.2.2.4 Topological Order of Rivers

The topological order of rivers, also called stream order or waterbody order, is a positive whole number used in geomorphology and hydrology to indicate the level of branches in a river system. Various approaches to the topological ordering of rivers or sections of rivers have been proposed based on their distances from the source or from the confluence (the point where two rivers merge) or river mouth, and their hierarchical positions within the river system. Topological ordering of the hierarchical relations of river patterns is consistent with multi-scale representation of spatial phenomena and is useful in river trees/networks description and

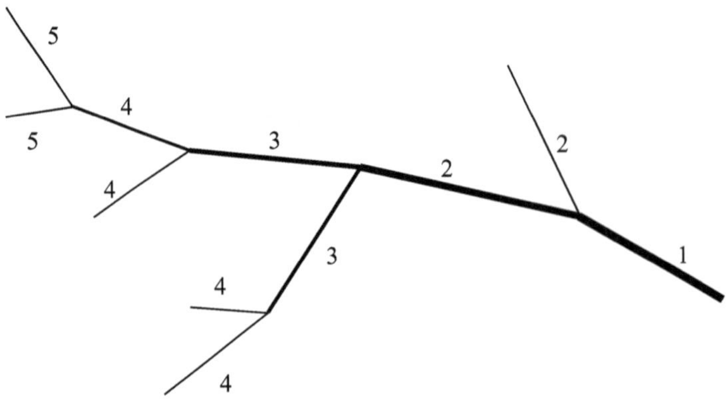

Fig. 5.7 Hack's stream order approach

generalization (mainly selection); thus, the following paragraphs present a number of approaches for ordering rivers.

Hack's Stream Order

Hack's stream order (Hack 1957) is also called the classic stream order or Gravelius' stream order. It is a bottom-up hierarchical ordering approach and works as follows (Fig. 5.7):

Step 1: the river segment (say, r) at the river's mouth at the sea (the main stem) is allocated the number a = 1.
Step 2: each direct tributary of r is assigned the number a+1.
Step 3: take each direct tributary of r as a new r and repeat Step 2 until all tributaries of the river are assigned numbers.

This stream order starting at the river's mouth indicates the river's place in the network. It is suitable for general cartographic purposes. It apparently can get the main stream of the river and easily get the topological relations among the river segments and main stream (Fig. 5.8).

Strahler's Stream Order

This top-down order system was devised by Strahler (1952, 1957). It works as follows:

Step 1: each of the outermost tributary is assigned the order number 1;
Step 2: when two streams with the order numbers b and c merge, if b = c,the resulting stream is given an order number b+1; If two streams with different order numbers merge, the resulting stream is given the higher of the two numbers.

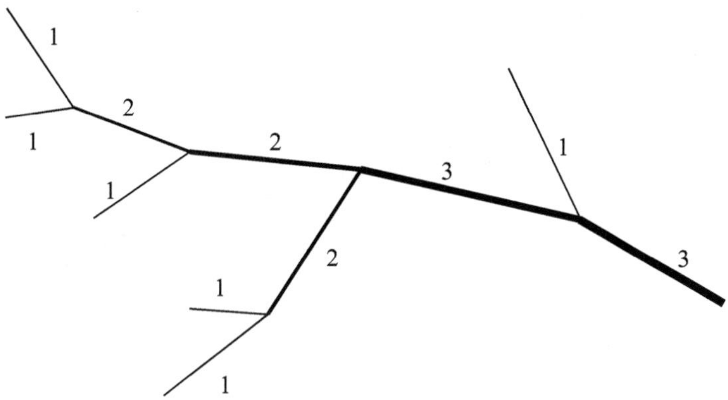

Fig. 5.8 Strahler's stream order approach

Step 3: recursively execute Step 2 until all river segments are assigned order numbers.

Strahler's order approach is designed for the morphology of a catchment and forms the basis of important hydrographical indicators of its structure, such as bifurcation ratio, drainage density and frequency. It is scale-dependent. The larger the map scale, the more orders of streams may be revealed. A general lower boundary for the definition of a "stream" may be set by defining its width at the mouth or, by reference to the map, by limiting its extent. The system itself is also usable for small-scale structures.

Shreve's Stream Order

The Shreve system is also a top-down order system proposed by Shreve (1966). It works as follows (Fig. 5.9):

Step 1: each of the outermost tributary is assigned the order number 1;
Step 2: when two streams with the order numbers b and c merge, the resulting stream is given an order number b+c.
Step 3: recursively execute Step 2 until all river segments are assigned order numbers.

Shreve's stream order approach is preferred in hydrodynamics: it sums the number of sources in each catchment above a stream gauge or outflow, and correlates roughly to the discharge volumes and pollution levels. Like the Strahler's method (Strahler 1952, 1957), This approach is dependent on the precision of the sources included, but less dependent on map scale.

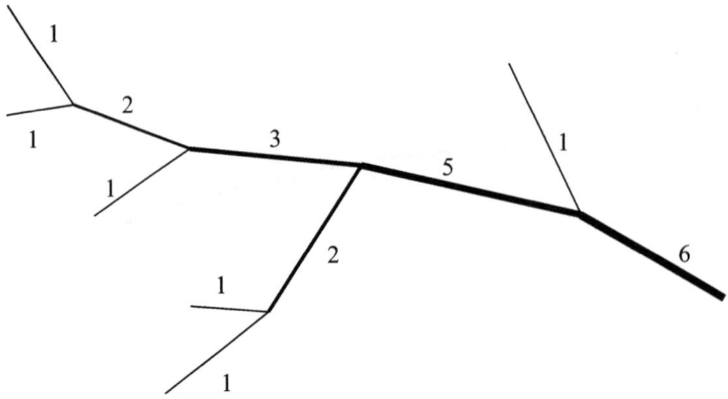

Fig. 5.9 Shreve's stream order approach

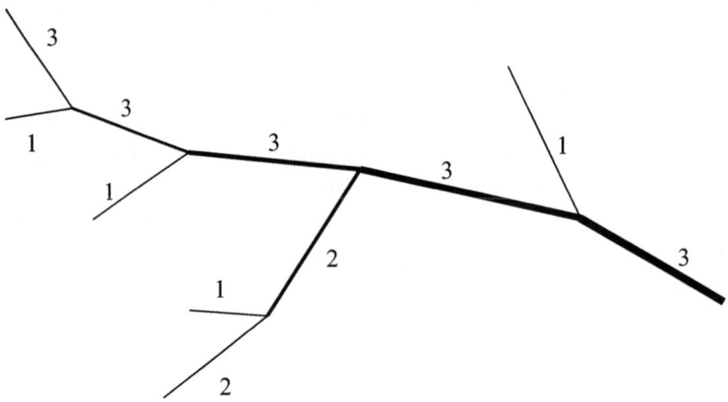

Fig. 5.10 Horton's stream order approach

Horton's Stream Order

Horton presented his approach to stream ordering (1945) based on the idea "the stream order is a measure of the position of a stream in the hierarchy of tributaries". As show in Fig. 5.10, in a given map of certain scale, the first order streams are those which have no tributaries, and the second order streams are those which have tributaries only first order streams and second order streams and so on. Horton's original system is somewhat more complex than this, in that, the streams of maximum order in the drainage basin is determined as in the extended back to its furthest source. In other words, the largest stream of the basin will be given the number of highest order.

Horton's scheme of ordering of rivers is difficult tedious and time consuming because it involves double phases of classification and re-classification at several times. During the second phase of renumbering some finger-tip tributaries are upgraded and other are left.

To sum up, Hack's ordering approach assigns a dimensionless numerical order (i.e. 1) starting at the mouth of a stream which is the lowest elevation point. The vector order then increases as it traces upstream and converges with other smaller streams, resulting in a correlation of higher order numbers to the higher elevation of headwaters (Hack 1957). Horton's idea based on vector geometry (1945) is a reversal of Hack's approach, and his approach is much complex in practice; thus, Strahler (1952, 1957) and Shreve (1966) proposed two modified versions to Horton's method. Both Horton and Strahler methods established the assignment of the lowest order (i.e. number 1) starting at the river's headwater, which is the highest elevation point. Whereas Hack's order number assignment correlates to height and elevation and traces upstream, Horton's, Strahler's and Shreve's stream ordering methods correlate to gravity flow and trace downstream, and they rely on principles of vector point-line geometry, and their rules form the basis of programming algorithms which interpret map data as queried by Geographic Information Systems.

5.2.2.5 Describing Rivers by the Graph Theory

Rivers on maps can generally be represented by connected lines and form linear networks. Because of the complexity of such networks, i.e. the approximations of drainage basins on the earth, their metric aspects such as length/width of river segments can be disregarded, the patterns of junctions may be paid more attention to in the communities of cartography and geographic information sciences.

If the metric aspects of the rivers on maps are ignored and only the connections and patterns are taken into accounts, one is focusing on "topology" of rivers. Because, in mathematical topology, the geometric figure of a drainage basin on the map consisting of patterns of connected lines on a plane (i.e. the map) forms a plane graph, the following section presents the method for applying the Graph Theory to represent drainage networks.

A plane graph consists of vertices and edges connecting the vertices. The edges can be either directed or nondirected. In a river network, the vertices are represented by the junction points and the initial points of the first-order streams. The edges are the river segments between junction points.

Assume that a plan graph (i.e. a river network) has n vertices (v_1,v_2,. . .,v_n) and m edges (e_1,e_2,. . .,e_m). It is possible to describe the graph by an associated matrix A with m × n elements. An element a_{ij} gives a quantitative expression from v_i vertex to vertex v_j.

Instead of the associated matrix, the graph can also be described by an incidence matrix B. The matrix element b_{ij} is

+1, if vertex v_i is the beginning of edge e_j;
−1, if vertex v_i is the end of edge e_j;
0, if vertex v_i is not on e_j.

The above two matrices can be used to describe the topology of river networks.

Except for islands and deltas, a river network is a special graph with no circuits. It has a "root" point (river mouth) and a number of pendant vertices (the beginning points of the lowest order river segments). Each edge is directed; from one pendant vertex to another there is one and only one path. Thus, such a river network, in mathematics, is an arborescence (or a tree). However, it should be noted that 'trees' are not necessarily representative of river systems because the flow direction is not specified on the edges and 'aqueducts' may occur (Scheidegger 1967).

5.3 Fundamental Principles for River Network Generalization

Besides a number of printed guidelines and regulations, the generalization of river networks/trees should abide by many rules and principles.

1. Original and target map scales are two key factors in map generalization. They determine to what extent the river networks should be simplified.
2. The purpose of the target map should be considered in the process of map generalization. The target maps of different purposes mean the maps are for different readers and/or users, and the rivers might be expressed differently.
3. The characteristics of mapped geographical region should be considered in which sometimes determines rivers selection rules and criterion. For example, rivers in southern China are very common and are regarded as ordinary natural features in the process of map generalization. However, in northern China's arid areas such as Gansu Province, rivers become rare and therefore precious and should be retained on the target maps as much as possible.
4. For the mapping areas with higher density of river networks, the criterions for river segments selection should be more rigorous than the areas with lower density of river networks so that the number of river segments selected can be moderately reduced; on the contrary, for the mapping areas with lower density of river networks, the criterions for river segments selection should be lowered so that the number of river segments selected can be moderately increased.
5. Density comparison of river segments among the mapping areas should be kept in the process of river segments selection, which means the areas with higher density of river segments should still have higher density before and after map generalization.
6. Although some special rivers do not meet given selection criterions, they still should be retained on the target map (e.g. a small river that flows into or out of a river, a small river that connects ponds or/and lakes, a small river that direct flows into the sea, a river in an arid area).
7. Although a river meets the selection criterions and its distance with a larger river on the target map is less than 3 mm, very closed, this river is generally deleted.

8. When the lines of the river are simplified, the basic shapes of the curvatures should not be changed.
9. Different criterions should be applied to generalize different types of drainage basins.
10. A river network should be kept as similar as possible before and after generalization.

5.4 Approaches for Generalizing River Networks

5.4.1 Selecting River Segments by Indices

Several models for river segments selection have been proposed, mainly including the linear regression model, the multiple regression model and the Radical Law (He 2004). They are conditional models, and each one has its advantages and disadvantages, and different models should be selected to generalize rivers in light of map scale, purpose of the resulting map, characteristics of the mapping area etc.

5.4.1.1 The Linear Regression Model

The basic index used in the linear regression model for river segments selection is density of river network. This index can be calculated by

$$\mathrm{D} = \frac{\sum L}{F} \tag{5.20}$$

Where,D is the density of the river network; $\sum L$ is the total length of the river segments in the river basin; F is the area of the river basin.

It is apparent in the geographic space that the greater the density of a river network, the greater the number of the rivers in the river basin. In other words, the density of a river network is correlated to the number of rivers:

$$N_0 = \frac{N}{F} \tag{5.21}$$

Where, N_0 is the number of rivers in unit area of the river basin; N is the number of rivers in the river basin.

It is easy to get the number of rivers of a river basin, the area of the river basin and the total length of the rivers in the river basin by the source map or the source database, then D and N_0 can be got, and criterions for selecting river segments in map generalization may be determined by them.

5.4.1.2 The Multiple Regression Model

The multiple regression model determines if a river segment can be retained on the resulting map by the number of rivers and the length of rivers in a unit area of the river basin. It may be expressed by a formula:

$$y = b_0 {x_1}^{b_1} {x_2}^{b_2} \tag{5.22}$$

where,y is the criterion for selecting river segments; x_1 is the number of rivers in a unit area on the source map; x_2 is the length of rivers in a unit area on the source map; and b_0, b_1 and b_2 are undetermined coefficients.

5.4.1.3 The Radical Law

The Radical Law or the Law of Selection (Töpfer and Pillewizer 1966) is also called the Square Root Model which gives a formula for calculating the number of features that should be retained on the resulting map:

$$n_F = n_A \sqrt{\left(\frac{M_A}{M_F}\right)^x} \tag{5.23}$$

Where, n_A is the number of the rivers on the source map; n_F is the number of the rivers on the resulting map; M_A is the denominator of the source map scale; M_F is the denominator of the resulting map scale; and x is a coefficient that can be calculated by

$$x = \frac{2 \ln \frac{n_F}{n_A}}{\ln \frac{M_F}{M_A}} \tag{5.24}$$

5.4.2 *Selecting River Segments by the River Tree*

Zhang (2006) proposed a river selection method using the river tree that has been discussed in a previous section of this chapter. This method can preserve density relations among different parts of a river network before and after map generalization.

Zhang's method (2006) is based on an improved rule for encoding river segments, which is called "branch rule":

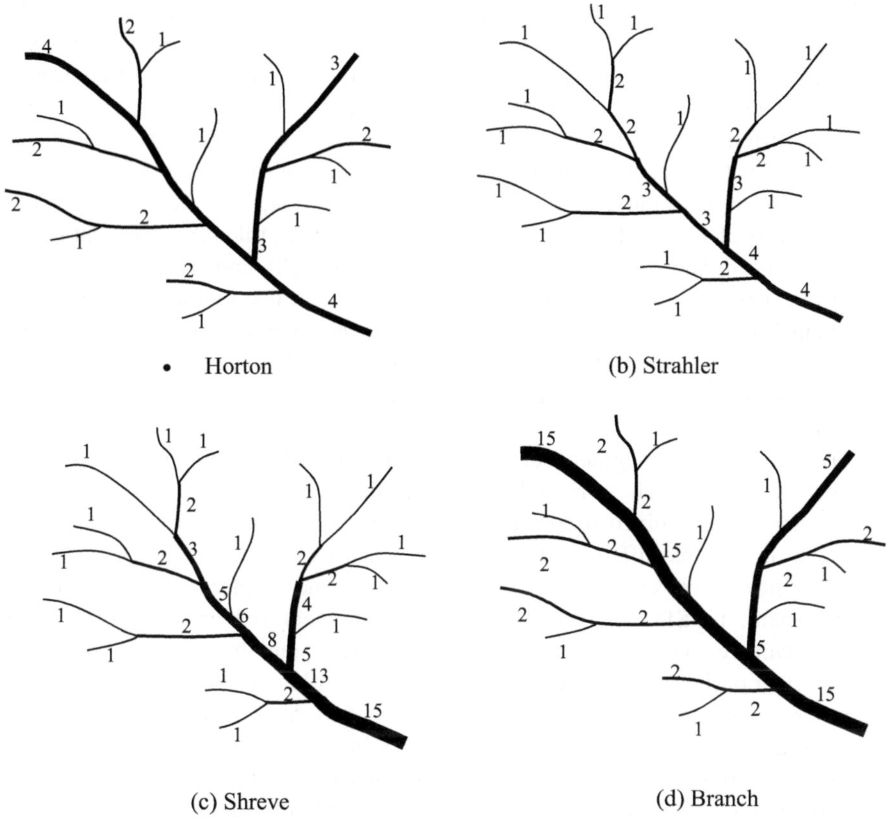

Fig. 5.11 Comparison among the four river ordering approaches (**a**) Horton (**b**) Strahler (**c**) Shreve (**d**) Branch

$$G = N_b + 1 \tag{5.25}$$

Where, G is the grade of a river R; and N_b is the total number of the branches of R.

Here, a river refers to a river entity but not a river segment. The branches include $R^{'s}$ branches at all grades but not only its direct branches. By this way, the more the branches a river have, the higher grade the river is. In this sense, the branch rule can tell grades of rivers. On the other hand, as far as a river with a specific length is concerned, the greater the number of the branches, the denser of the river network. In this sense, the branch rule can tell density of river networks.

The reason that the branch rule is used in river selection is because it has the following advantages compared with other rules including the rules proposed by Horton (1945), Strahler (1952, 1957), and Shreve (1966) (Fig. 5.11). Horton's and Strahler's orders express the hierarchy of river entities or river segments and can tell the position of each river entity or river segment in the river tree. Horton's approach takes river entities as units and Strahler's approach takes river segments as units in ordering river networks. Horton's and Strahler's approaches do not increase the grade of the new river segment/entity when two river segments/entities merge, and they can tell neither the differences of river network densities nor the differences of the branch numbers. Although Shreve's ordering approach constructs river trees and tells the differences of the branch numbers belonging to each river segment, the differences of the branch number are at the beginning of the river segment but not between the origin of the river segment and the river mouth; thus, the ordering results by this approach do not show the river density. The branch rule takes river entities as the units in ordering (this is similar to Horton's approach), and when two branches merge, the grade of their main stream increases (this is similar to Shreve's approach). The branch rule tells the total number of the branches merged into the main stream from the river origin to the river mouth, and the differences of the number of branches in unit distance can indicate the differences of the density of the river network.

Traditionally, river segment selection is a complicated process which considers a number of factors such as river length, river density, and river grade, and these factors are usually mixed used which makes design and implementation of river selection algorithms difficult. Hence, if these factors can be combined to form a complex index, river selection algorithms might become simpler to design and easier to implement. It is a fact in manual cartography that rivers are hierarchically selected, i.e. the mainstream of the river basin is selected firstly, and then the mainstreams of the sub-river basins are selected, and so on. Here, river length is regarded as the most important factor, and density of river network and river grade are regarded as assistant factors, and they are used interactively in river selection.

To simple this situation in river selection algorithms, Zhang (2006) proposed a complex index which integrates the above factors into a formula:

$$\mathrm{I} = \mathrm{L}\left(\frac{G}{L}\right)^{\alpha} \times (H_m - h + 1)^{\beta} \qquad (5.26)$$

Where,I is the complex index of the river;L is the length of the river; G is the grade of the river; H_m is the maximum hierarchy of the river; h is the number indicating which hierarchy the river is at; and α and β are two parameters determined by cartographers' experiences.$\alpha \epsilon [0, 1]$; and $\beta \epsilon [0, 1]$.When $\alpha = \beta = 0$, the formula becomes $\mathrm{I} = \mathrm{L}$.

This formula manifests that a river with greater length, higher density and greater grade have priority to be selected.

If L,*G*and *h* of the river are given and do not change, cartographers generally hope the complex index (i.e. I) can increase with the increase of α and β. However, in practice, if $\frac{G}{L} < 1$, the complex index decreases with the increase of α. To improve this situation, the complex index can be calculated by the following formula:

$$I = L \times G^{\alpha} \times (H_m - h + 1)^{\beta} \tag{5.27}$$

To get the complex index I of a river by Formula 5.27, the length, the grade and the hierarchy of the river should be got firstly. Because it is easy to get the length of the river, only a method for obtaining the other two factors are discussed here, proposed by Zhang (2006), including three steps.

Step 1 data preprocessing: above all, the closed loops in the river network are marked and deleted so that the network becomes a pure tree structure; and then the river mouth is marked (a river tree has only one river mouth and thus the mouth can be easily obtained and marked).

Step 2: construction of river entities: search river segments from the river origin to the downstream. When several river segments meet at a junction, the longest river segment is viewed as a part of the mainstream (i.e. a section of the river entity) and the mainstream can be extended to the downstream until all segments belong to the mainstream are found and connected which means the river entity is formed.

However, the longest river segments are not always regarded as the mainstream in many cases in the geographic space because of historical reasons; thus, if the river segments have names their names should be taken into consideration and the river segments with the same names can be connected to form a river entity.

Step 3: formation of the river tree: the mainstream and its tributaries may be recognized and recorded in the process of constructing river entities, hence the river tree is formed naturally. In the meanwhile, the length, grade and the hierarchical position of each river segment are calculated and saved.

After the above three steps, the complex indices of the rivers can be obtained and by which the river entities that can appear on the resulting map may be selected.

5.4.3 A Knowledge-Based Approach to River Network Generalization

Artificial intelligence has been a tool used in map generalization by many researchers (Li et al. 2010) basically abiding by the following ideas: extract and express the knowledge and rules used in the process of map generalization, and then spatial reasoning methods are employed to generalize (e.g. select, simplify, collapse, move) map features.

Here introduces a knowledge-based approach to river network generalization proposed by Wu et al. (2007). This method classifies the knowledge used in river network selection into three categories, i.e. spatial knowledge, attribute knowledge and river generalization rules, and studies on how to obtain and formally represent such knowledge and constructs a rule library for river network selection. This rule library can be used in carrying out river selection by knowledge inference, and it considers the spatial and attribute knowledge of the river network and the priorities and weights of river generalization rules.

5.4.3.1 Acquiring Knowledge for River Network Selection

There are mainly three kinds of knowledge used in river generalization: the knowledge contained in the attributes of rivers, the knowledge regarding to spatial relations among river segments, and the fuzzy knowledge contained in some qualitative factors such as river grade and the hierarchical position of the river segment in the river network.

Firstly, the original river data are processed to get a river tree. Generally, the original river data are saved as separated singular lines in the spatial databases. To get a river tree, they are processed in light of the graph theory. The spatial topological relations among the river segments (i.e. the singular lines) are calculated and are saved in so-called RIVER-ARC and RIVER-NODE structures. The RIVER-ARC structure capsules the directions and the lengths of the river segments and topological relations (e.g. joint relation, neighbouring relation) among river segments; RIVER-NODE structure capsules the coordinates of the vertices of each river segment. The two structures lay a data model foundation for constructing river networks, because by which the river data are re-organized and the river entities are constructed which contain spatial knowledge of the river.

Secondly, the rules for river network generalization are represented, which consist of the knowledge at three aspects: (1) the rules and indices for river selection from mapping specifications, (2) the graphic patterns and characteristics of the river, and (3) the weights of the river.

It is easy to get the rules and indices from mapping specifications; however, use of these rules in map generalization is difficult, because the rules and indices from mapping specifications are usually expressed using constant values and cannot be directly used in the generalization of the maps from different geographic spaces. Obtaining the graphic patterns and characteristics of the river usually can be reached by the graph theory, pattern recognition and artificial intelligence, though it is difficult to achieve the goal. As far as the weights of the rivers are concerned, the rivers owning the following three characteristics can be assigned greater weights and therefore have priority to be retained in map generalization:

- the rivers at the edge of the river basin: construct the river tree for the river basin. The tributaries at the first and the last level of the river tree belong to the rivers at the edge of the river basin.
- the rivers connecting with lakes or flowings into seas: they can be easily judged and obtained by the spatial relations among the rivers, lakes and seas.
- the rivers with higher grade, flowing past more residential areas and/or at the border of political regions should be assigned greater weights and have priority to be retained on the resulting map in map generalization. The grades of the rivers

can be got by the river basin's tree structure; the residential area of the river may be obtained by buffering the river; and whether a river is at the border of political regions can be judged by the topological relations among the river and the regions.

Lastly, structural knowledge of the river basin is extracted. The river basin can be viewed as a river tree consisting of mainstreams at different levels, i.e. mainstreams at first level, mainstreams at the second level etc. The tributaries connecting with the river mouth are called the mainstreams at the last level. The structure of a river basin can be organized in this way which makes the construction process of river trees concise, simple and reasonable.

After getting the river trees, some useful information regarding rivers should be obtained and attached to the rivers, including river flow direction, river grade, river length, distance between adjacent tributaries, relations of a tributary with its mainstream (i.e. if the tributary is the left or right child of the mainstream) etc.

5.4.3.2 Generalization of Rivers

The rule library can be established by the knowledge for automated river generalization. Each rule in the rule library is expressed by a function and the procedure regarding how to use these rules is designed, expressed and saved in the library. Obviously, this inference procedure determines river selection results; therefore, the river knowledge, river selection rules and their weights and priorities in the rule library should be considered jointly. Here, a forward reasoning method is designed which starts from the facts and known data to get resulting maps by the rule library. In the process of river selection, all of the rivers that can trigger the inference procedure should be selected firstly so that the procedure may be started, and then new facts and rules are selected to push the inference procedure forwards until each of the rivers is judged (Fig. 5.12).

If a river basin is generalized using the methods based on spatial knowledge reasoning, two points need to be noticed: if the knowledge is correct and if the inference procedure is reasonable. Here presents an example (see Fig. 5.13) for selecting rivers by reasoning.

Step 1: if AB is the mainstream of the river who at least has one tributary and the tributary has no tributary (e.g. CD, EL), select a tributary who has no tributary (i.e. CD in this example). Then, search the river tree. If no tributary who has no tributary can be found, end this procedure.

Step 2: judge if the river (i.e. CD) can be retained by some of its important attributes such as if it is a perennial river. Further, select some quantitative indices such as river length and river interval to judge if the river can be retained.

Step 3: judge if the river length and river interval of CD meet the indices for retaining it on the resulting map. If they do, retain this river and go to Step 1; otherwise go to Step 4.

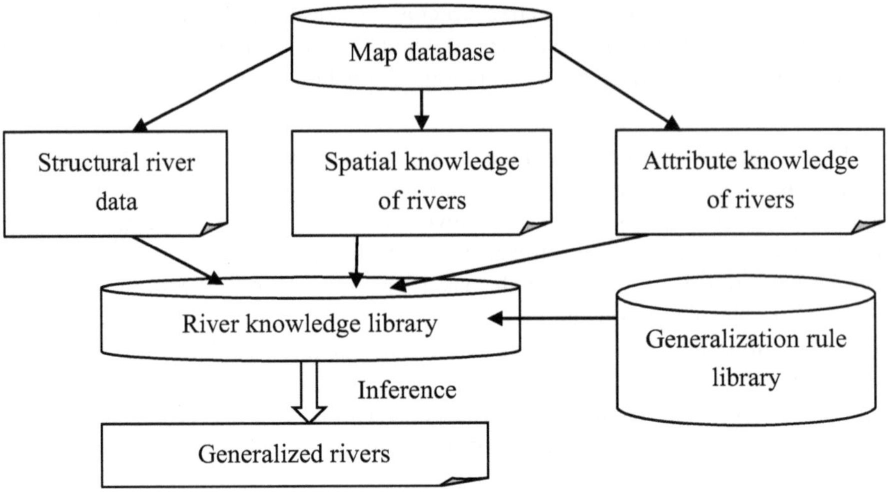

Fig. 5.12 River generalization by rules. (After Wu et al. 2007)

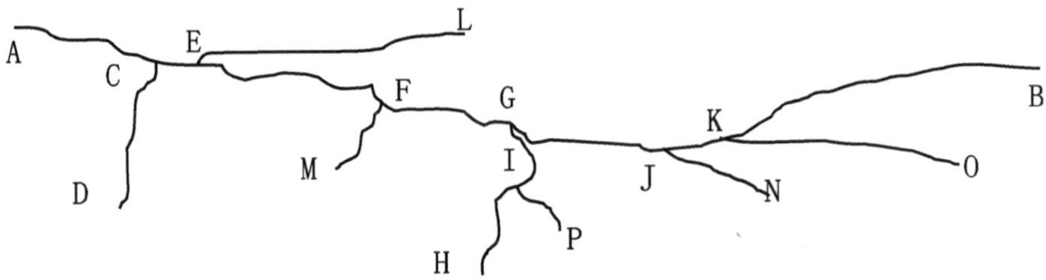

Fig. 5.13 River selection process by rules. (After Wu et al. 2007)

Step 4: judge if this is an important river by the three rules mentioned previously: (1) if the river at the edge of the river basin, or (2) if the river connects with lakes or flows into seas, or (3) if the river has higher grade or flows past more residential areas or is at the border of political regions. If it meets one of the three rules, it can be retained on the resulting map. Otherwise, mark this river as "will-be-deleted" and go to Step 1.

A river marked as "will-be-deleted" is not certainly deleted. Whether it will be physically deleted from the database depends on other factors such as graphic characteristics of the river basin. For example, in Fig. 5.13, tributary EL does not meet the basic rule for retaining it on the resulting map, nor is it an important river. However, it is the only right tributary of the river tree, therefore it should be retained for the purpose of keeping graphic similarity of the river tree before and after map generalization.

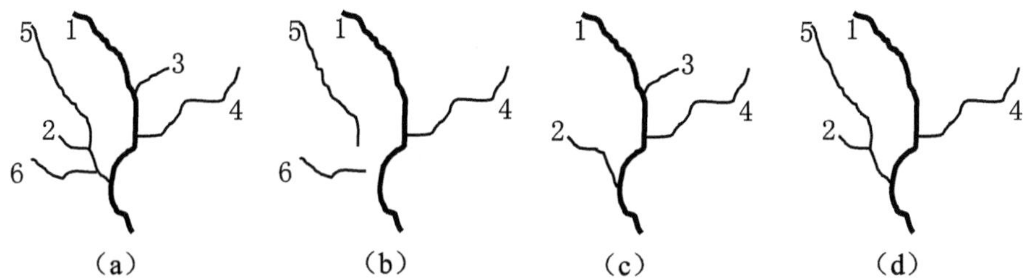

Fig. 5.14 Use of different factors generate different results (After Ai et al. 2007)

5.4.4 An Approach to Generalizing River Networks by Catchment Area

Grade, length, catchment area and other attributes of rivers are usually considered in river network generalization (Ai et al. 2007). These factors are correlative to each other, and their correlative degrees generally change when geographic characteristics of the mapped area change; thus, it is difficult to use such factors in river network generalization.

If only one factor is considered the result is usually wrong or/and unacceptable. So, cartographers generally use multiple factors in river network generalization. The following three examples give an explanation of this issue, taking Fig. 5.14a as the original river network: on which 1 is the mainstream, and 2, 3 and 4 are the direct tributaries of the mainstream, and 5 and 6 are the direct tributary of 2.

Example 1: If only river length is taken as the criterion and a threshold of river length is given and used in river selection, tributary 2 might be deleted and tributaries 5 and 6 might be retained (Fig. 5.14b). Nevertheless, this result is unacceptable.

Example 2: if only river grade is used in river selection, the rivers 5 and 6 are deleted and the other ones are retained on the resulting map, shown in Fig. 5.14c. This result is unreason, because tributary 5 is very long and is usually regarded as an important tributary; hence, it should be retained on the resulting map.

Example 3: if both river length and river grade are considered, the resulting map might be Fig. 5.14d: tributary 5 whose grade is low but whose length is much longer than many other tributaries is retained while tributary 3 whose grade is high but whose length is much shorter is deleted. Apparently, this result is more reasonable than the other two.

If river catchment area is used as a criterion in river selection, results can be improved (Ai et al. 2007), because on one hand, it is reasonable that the greater the catchment area of a river, the more probably the river can be retained on the resulting map; on the other hand, the hierarchical relations among the polygons of the catchment areas can ensure that the catchment area of a river is greater than that of its any tributary. Hence, the case in Fig. 5.14b is impossible if river catchment area is taken into consideration in river selection.

In addition, values of river catchment areas depend on the spatial distribution of the river segments. For example, there are two tributaries A and B in a river tree, and

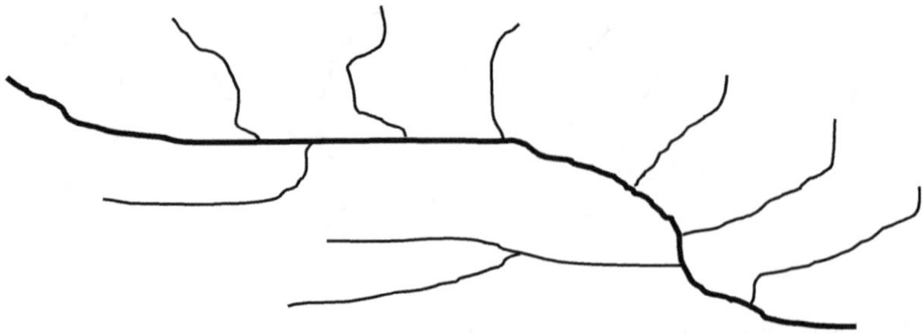

Fig. 5.15 Tributaries are unevenly distributed at the two sides of the mainstream

their lengths are equal. Tributary A has higher grade and is in an area of higher river network density; while tributary B has lower grade and is in an area of lower river network density. Under this condition, the catchment area of A is generally less than that of B. Thus, if river catchment area is used in river selection, a tributary with lower grade but greater catchment area has greater probability to be retained on the resulting map than the one with higher grade but less catchment area (see Fig. 5.14c as an example). Such case is very common in the region where river basins are extremely unevenly developed. If rivers selection only considers river grade then graphic characteristics of the river networks will be distorted.

The above analysis tells that the catchment area is a key index for evaluating the degrees of importance of rivers. However, in some cases, the catchment areas of tributaries have little differences, therefore it is inappropriate to use catchment areas to evaluate the importance of tributaries. For example, in Fig. 5.15, if catchment area is employed as the criterion, the six tributaries at the northern side of the mainstream will either totally deleted or totally retained, which is obviously unreasonable. To overcome this disadvantage, a new method using catchment area index has been proposed (Ai et al. 2007). This method includes two steps:

Step 1: take the interval between tributaries as the index in river selection and do initial selection. To achieve this goal, the tributaries at the leaves of the river tree are triangulated, by which the distance between a tributary and the tributaries at the tributary's two sides are calculated respectively. If a distance value is less than human's visual discrimination distance on the resulting map, the corresponding tributary is deleted.

Step 2: calculate the number of river segments that can be retained on the resulting map by the Radical Law (Töpfer and Pillewizer 1966), and then sort the tributaries in decreasing order according to their catchment areas. When the tributaries are selected, retain the tributaries one by one until the number of retained tributaries are equal to the number calculated by the Radical Law.

In this method, after execution of Step 1, the tributaries at the area of very high density of river networks are deleted, and the catchment areas occupied by these tributaries are divided by other adjacent tributaries, which naturally increases the probabilities of the retained ones to be retained on the resulting map when Step 2 is executed.

The method takes into consideration the grades, lengths and catchment areas of tributaries and the intervals among tributaries, and it regards the catchment area as an integration of the grade and length of the tributary and the intervals of the tributary with its neighbours, which makes river selection process simple and the results reasonable.

5.4.5 River Selection by the BP Neural Network Algorithm

Back propagation (i.e. BP) is a method used in monitor learning (Ethem 2010) and artificial neural networks to calculate a gradient that is needed in the calculation of the weights to be used in the network. It is commonly used to train deep neural networks, a term referring to neural networks with more than one hidden layer. Back propagation is a special case of a traditional and more general technique called automatic differentiation. In the context of learning, back propagation is commonly used by the gradient descent optimization algorithm to adjust the weights of neurons by calculating the gradient of the loss function. This technique is also sometimes called backward propagation of errors, because the error is calculated at the output and distributed back through the network layers. The BP learning process consists of two steps:(1) Forward propagation of operating signal, and (2) Back propagation of error signal.

The back propagation algorithm has been repeatedly rediscovered and is equivalent to automatic differentiation in reverse accumulation mode. Back propagation requires the derivative of the loss function with respect to the network output to be known, which typically means that a desired target value is known. For this reason, it is considered to be a supervised learning method, although it is used in some unsupervised networks such as autoencoders. Back propagation is also a generalization of the delta rule to multi-layered feed forward networks, made possible by using the chain rule to iteratively compute gradients for each layer. Because a river basin can be represented by trees or networks, and its generalization process is similar to the basic idea of the back propagation algorithm, some scholars have used it in automated selection of river networks. Here presents an algorithm proposed by Shao et al. (2004) which consists of three steps: organizing river data, acquiring leaf nodes, and selecting rivers by the BP neural network technique.

5.4.5.1 Organizing River Data

Take the river as the basic unit and the river data are organized using river data structure: firstly, the mainstream of the river basin is constructed according to the methods mentioned in the previous sections of this chapter; secondly, hierarchical relations of the tributaries are obtained and a river tree is formed. Lastly, the

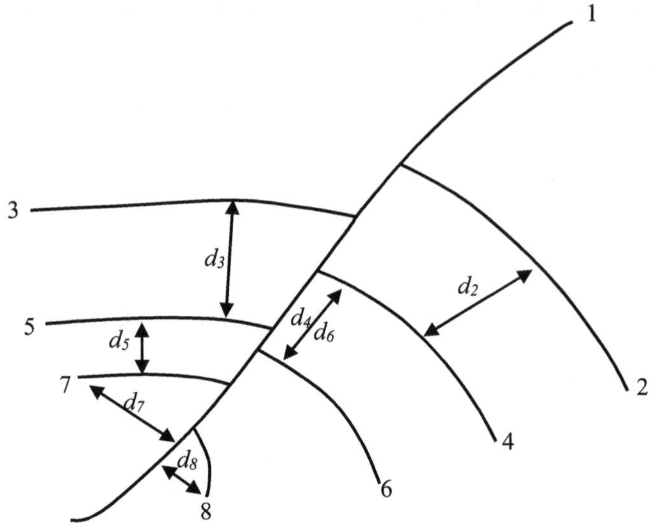

Fig. 5.16 River intervals

attributes of each river segment are calculated and assigned to the river segment, and the hierarchical relations among the mainstream and the tributaries as well as their attributes are saved in a 2-dimensional table.

5.4.5.2 Acquiring Leaf Nodes

At this step the length of each river corresponding to a leaf node, the interval of two rivers corresponding two leaf nodes and quantitative importance of each river corresponding to a leaf node are calculated. Because it is easy to get the length of each river, only the other two aspects are discussed here.

5.4.5.3 Calculation of River Interval

Figure 5.16 shows an example of river interval calculation. Three cases should be considered.

Case 1: if the river is the first leaf node in the river tree

$$\mathrm{d} = \mathrm{Min}\ \mathrm{d}(\mathrm{a}, b_i), b_i \in \mathrm{B} \tag{5.28}$$

where, d(x, y)is the distance between x and y; a is the starting point of the current river; and B is the set of the direct downstream of the current river.

Case 2: if the river is at the middle of the leaf nodes of the river tree

$$d_{front} = \text{Min d}(a_i, x_i), a_i \in \text{A}, x_i \in \text{X}$$
$$d_{back} = \text{Min d}(a_i, y_i), a_i \in \text{A}, y_i \in \text{Y}$$
$$\text{d} = \text{Min}(d_{front}, d_{back}) \tag{5.29}$$

where, A is the set of the rivers corresponding to the leaf nodes of the river tree; X is the set of the rivers at the same level of A and at the same side of A and in front of A and each of the rivers corresponding to a leaf node of the river tree; and Y is the set of the rivers at the same level of A and at the same side of A and behind A and each of the rivers corresponding to a leaf node of the river tree.

Case 3: if the river is the last leaf node in the river tree

$$\text{d} = \text{Min d}(a_i, x_i), a_i \in \text{A}, x_i \in \text{X} \tag{5.30}$$

where, A is the set of the rivers corresponding to the leaf nodes of the river tree; and X is the set of the rivers at the same level of A and at the same side of A andneighbouring A and each of the rivers corresponding to a leaf node of the river tree.

5.4.5.4 Acquirement of Importance Degrees of Rivers

The importance degree of a river is complex concept, because it is related to multiple factors. Here, those geographically special rivers that connect with ponds, lakes and seas, or the borders of political regions, or perennial ones in arid areas are not discussed, but only common rivers are considered, and a formula for calculating their importance degrees are given:

$$\text{p} = \alpha x_1 + \beta x_2 \tag{5.31}$$

where, the residential areas are classified into two levels; p is the importance degree of the river; x_1 is the residential areas at the first level; x_2 is the residential areas at the second level; α is the weight of the residential areas at the first level; and β is the weight of the residential areas at the second level.

It should be noticed that to ensure $\text{p} \in [0, 1]$,x_1,x_2,α and β need to be normalized before they are used in the calculation of p.

5.4.5.5 Selecting Rivers by the BP Neural Network Technique

The algorithm proposed by Shao et al. (2004) considers three factors in the process of river selection, i.e. river length, river interval and river importance degree, and the result is the selected rivers; thus, in this algorithm the BP neural network has three

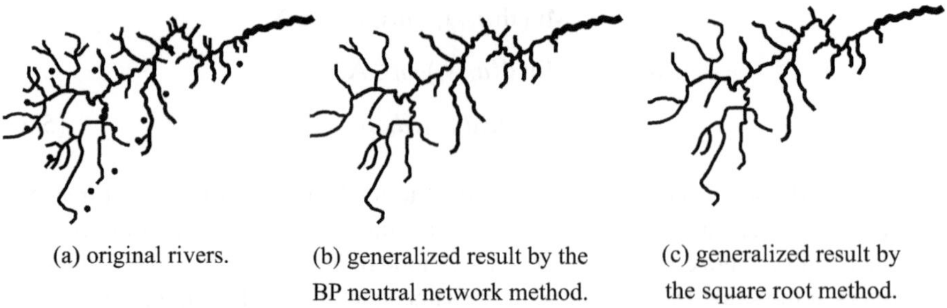

Fig. 5.17 Test of the BP neutral network method (**a**) original rivers (**b**) generalized result by the BP neutral network method (**c**) generalized result by the square root method

neurons at the input layer of the network and one neuron at the output layer of the network.

The data used in the machine learning are from a topographic map at scale 1:1 500 K. The rivers and buildings are extracted and re-organized, and the river lengths, river intervals and river importance degrees are calculated and saved in a database. The rivers and buildings on the map are generalized by experienced cartographers to get a map at smaller scale 1: 3 M. Then the original map and generalized map are used as sample data for training the BP neural network algorithm, and finally the weight matrix and improved algorithm can be achieved.

Figure 5.17 shows an example for testing the BP neural network algorithm proposed by Shao et al. (2004). The resulting river network generated by this algorithm is better than the square root algorithm, because it takes the weights of the rivers into consideration and uses the BP neural network technique which may make the algorithm smarter by accumulating map generalization through machine learning.

5.4.6 *River Network Selection Based on Structure and Pattern Recognition*

Generalization aims at reducing the level of details of a database in order to meet new specifications, which includes cartographic generalization and model generalization. The former aims at producing maps and the latter aims at generating new databases of source data. As far as river network generalization is concerned, a lot of achievements have been made for many years, and most of them are on cartographic generalization. Thus, our concerns move from the former to the latter. Here introduces an algorithm proposed by Touya (2006) which addresses database generalization, specifically river network selection. The algorithm uses the principle of "good continuation" to enrich the database by constructing so-called "river strokes" and selects river networks based on river structures and patterns. It mainly consists of the following steps.

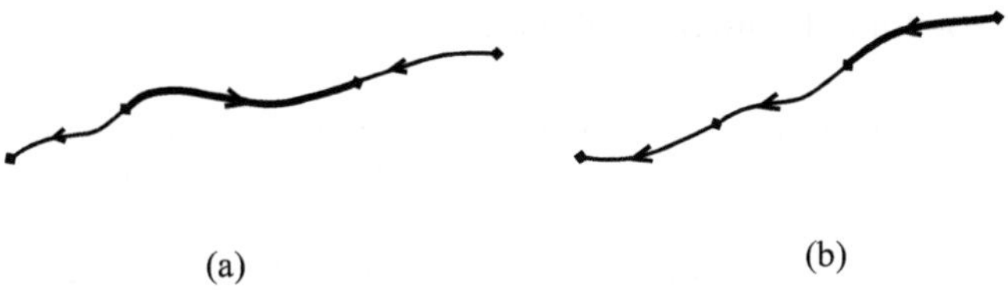

Fig. 5.18 Two examples of wrong flow directions: (a) the lightened segment has a wrong flow direction as it is inconsistent with its neighbours; and (b) the lightened segment has a wrong flow direction, inconsistent with its neighbours, and it is the first river segment from a source

5.4.6.1 Data Pre-processing

River data from map database is not always correct. For example, the flow direction of the river segments in some cases is not completely reliable. So, the initial dataset has often to be cleaned to correct the errors in flow direction. Two processes are successively applied to infer flow directions in the river network.

The first process: a process adapted to initial data is carried out. It consists in using elevation data of stream segments (i.e. attribute data). Each stream segment has an elevation value for its initial vertex and final vertex. Obviously, if the elevation value of final vertex is greater than elevation value of initial vertex, stream segment geometry has to be reversed to be consistent with the flow direction.

The second process: this process is more independent to initial data as it consists in a neighbourhood analysis. Locally analyzing the inflows and the outflows of streams may allow to infer flow direction for simple cases like in Fig. 5.18. At sources or sinks, neighbourhood analysis may correct other errors of flow direction.

Furthermore, generalization processes often need topologically correct network data to provide meaningful results. Most real data contain errors and especially topological errors. So, a check of the topological relations among river networks is necessary. After these pre-processing steps, the selection process itself can take place.

5.4.6.2 Enriched Data Schema for Generalization

Automatic selection of river networks is a complex process that often needs to enrich the raw dataset to be generalized to recognize implicit structures and patterns.

To implement this algorithm in an object-oriented programming language (e.g. C++), a "River stroke" class is added to store the strokes that is used for selection. The classes "Source" and "Sink" are added to store the beginning and the end of the strokes so as to use them as database objects in further processes. Two classes are also added to manage the selection of river islands. In addition, a class to store irrigation zones is added. Indeed, the strokes building procedure does network in these zones, so they have to be previously detected.

5.4.6.3 Extraction of Sources and Sinks

In order to build the "river strokes" and the irrigation zones, it is firstly necessary to build the sources and sinks. A source is a node of the network that has a single output link and no input link. A sink is the opposite: a node with a single input link and no output link. Sources and sinks can usually be extracted from the river data by their attributes such as "type".

5.4.6.4 Construction of River Strokes

The strokes building algorithm is a downstream pass: it begins at the sources of the network and ends when all the nodes (sources, sinks and confluences points) have been treated. "River strokes" are built to correspond to the classic ordering of rivers: a river stroke begins at a source and ends at a sink or at a confluence point with a more important river stroke. The tricky part of the creation of such strokes is to define the main stream at a confluence point. To get more realistic results, the Gestalt principle of "good continuation" is constrained by rules and the river strokes which are different from the purely geometric perceptual strokes. Many rules can be used to define the main path. Horton (1945) used the longest and straightest path, and Thomson and Brooks (2000) used the river name, the longest and straightest path, others used the largest drainage basin. The rules used here are a bit more complex:

1. Strokes always follow a named stream.
2. All other things being equal, a "permanent" regimen river has priority on an "intermittent" regimen river.
3. All other things being equal and the sources of the streams being "natural", the longest and straightest path has priority. The length is the major parameter because it is more relevant than straightness. Straightness is rather used when the confluence angle is greater than a threshold (60°) and if the length difference is less than a threshold (500 m).
4. All other things being equal and one of the sources of the streams being "zone limit", the straightest path has the priority.

To follow the good continuation principle, straightness means curvature continuity. The straightest path is the one that leads to more curvature continuity with the upstream stroke. The length of a river is measured from the source point to the confluence point.

River networks are sometimes composed of deltas and islands where a stream splits into two or more channels in the downstream direction. So, two types of river strokes are distinguished: "main channels" and "braided streams". When a stream splits into two or more channels, one outflow is considered as the main channel and continues the "main channel" stroke and the others are used as the first segment of new "braided" river strokes. The chosen main channel is the straightest path (Fig. 5.19). The shortest path to the sink could have been used instead but the

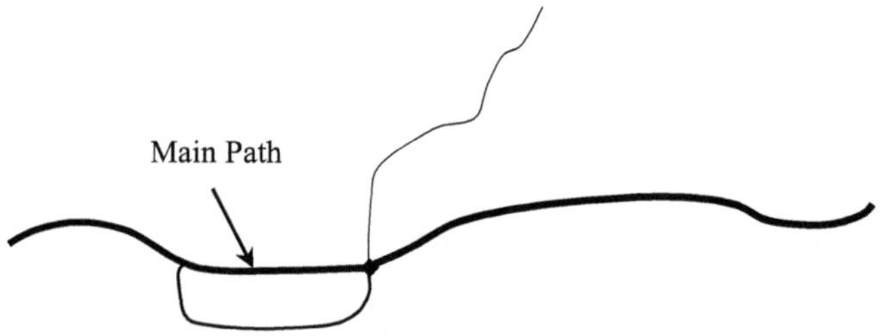

Fig. 5.19 The main stream on a confluence point with two outflow and two inflow streams

process would have slowed down as the shortest path is a computation time consuming algorithm and the contribution is not sufficient. But this rule could be used for small datasets.

5.4.6.5 Formation of Islands

In river networks, braided streams often correspond to islands on the rivers, and there are often several adjacent islands on the river. These islands are significant data and it is interesting to select the outline of some islands in the generalized dataset. In the process of the data enrichment, the aim is to form islands and complex islands as database objects to make the selection of the outline easier. A complex island is an aggregation of islands. In the enriched dataset, each island is linked to the stream segments of its outline, even for complex islands. If selection specifications require a different level of abstraction for islands and especially a smaller level, adjacent islands can be aggregated to several adjacent complex islands in order to keep the information of adjacent islands after the selection. Then, the enriched data model would be a bit modified to allow adjacent complex islands.

To build such islands, the topologic faces of the network that are regarded as a graph have to be built. Only the small faces, the ones that really represent islands, are kept to give the geometry to the new objects: a threshold is determined based on the test dataset. Complex islands are built by clustering the simple islands. In order to build adjacent complex islands for a smaller level of abstraction, a hierarchical clustering would rather be used.

5.4.6.6 Detection of Streams Inside Irrigation Zones

The process for detecting irrigation zones is necessary, because irrigation zones are sources of errors during the creation of river strokes. As explained in previous sections of this chapter, the stroke creation procedure is based on flow directions, and in irrigation zones most streams are artificial and flat and have no real flow direction, because the river segments are often given an arbitrary flow direction in the original data. The inconsistency of flow directions obviously may forbid the algorithm to go further.

Irrigation zones are characterized by a flat ground, a strong density of straight and short streams and many sources and sinks. These characteristics are used to automatically detect irrigation areas and to build them as objects in the dataset. Then, compactness is used to remove over-detected areas.

The next step is the automatic recognition of artificial streams in regards to natural ones within irrigation zones in order to remove them from the selection process. Artificial streams are rather short and straight whereas natural ones are rather long, sinuous and have many intersections with other streams. To translate the difference into geo-computational measures, the strokes are used once again but this time with geometrical strokes (only curvature variation is used). So, strokes are computed in each irrigation area and then characterized by measures like sinuosity and number of intersections with other strokes in order to be differentiated.

The detection of natural and artificial streams inside irrigation zones ends up the data enrichment step of the proposed river network selection process.

5.4.6.7 River Selection

After completion of the data enrichment step, the river selection step itself can take place. The major part is the selection of the river strokes. As strokes are built to represent whole rivers, selection rather concerns stokes than segments. All segments of a stroke are selected or none is. In order to determine which strokes are to be selected, the main criterion used is a hierarchical organization of the strokes.

The Horton ordering (Horton 1945) is applied here on organizing strokes. To compute the Horton ordering, a Strahler ordering of the river segments is needed and used. The Strahler order is assigned to the river segments during the strokes building procedure. When a segment extends a river stroke, its Strahler order is computed. The segments belonging to a "braided stream" stroke are not used to compute the Strahler order and no Strahler order is assigned to them as their selection is not based on their own hierarchy. The Horton order of each stroke is finally computed after the building of all strokes in the network: the Horton order of a stroke stream is the maximum of the Strahler orders of the river segments that compose it.

The selection process is based on the river strokes using the Horton order to organize them in hierarchy. The less important ones in the hierarchy are not selected.

The selection criteria are a threshold on the Horton order of the strokes and a threshold on the stroke length. The river strokes are selected according to their type: "primary" or" braided". High Horton order primary strokes are always selected and low ordered primary strokes are selected if they are long enough. The braided strokes which are long enough are selected only when their primary stroke is selected. All the primary strokes with a Horton order of three or more are selected. The strokes with an order of two or one are selected if the length is longer than specific values. The braided strokes are selected if their primary stroke is selected and if their length is longer than a specific value. Obviously, the parameters may be different in different cases.

To conclude, this algorithm is based on previous work on river network generalization, especially that introduces the notion of "strokes" but also for the organization in hierarchy of the streams of the network. This work adds the management of river islands, irrigation zones and allows the building of strokes on a clipped area where some sources are not natural. A data correction step dealing with network topology and flow direction has also been developed. The complete selection process is composed of three steps: data correction, enrichment and selection.

5.4.7 Comparison of the Algorithms for River Network Generalization

Automated map generalization is always an important issue and a major challenge in cartography and Geographical Information System research. Regarded as the skeleton of the terrain, the drainage system is already considered to preserve terrain features in generalization. By far many researchers have paid attention to the generalization of river networks and many approaches for river selection have been developed, typically including the approach for selecting river segments by indices (He 2004), the approach for selecting river segments by the river tree (Zhang 2006), a knowledge-based approach to river network generalization (Wu et al. 2007), an approach to generalizing river networks by catchment area (Ai et al. 2007), an approach to river selection by the BP neural network algorithm (Shao et al. 2004), an algorithm for river network selection based on structure and pattern recognition (Touya 2006), etc. Obviously, each approach has its advantages and disadvantages and each can be used in different cases. To use them efficiently, these approaches are compared and the results are shown in Table 5.1.

5.5 Summary of the Chapter

River network is a type of important features on maps and in map databases. Multi-scale representation of river networks is necessary in constructing spatial data infrastructures.

This chapter introduces two categories of methods for describing river networks that can be found in literature. The first one is to describe river networks using measures and parameters, and the second one is by special methods for specific purposes including the information entropy-based method, the fractal dimension-based method, the river tree method, the topological order method and the graph theory-based method. These methods lay foundation for automated generalization of river networks on maps. Then it addresses the Fundamental principles that should be

Table 5.1 Comparison of river network generalization approaches

Approach name	Preserve density relations	Preserve graphic similarity relations	Preserve river patterns	Generalized results
Select river segments by indices	Ordinary	Ordinary	Ordinary	Ordinary
Select river segments by the river tree	Good	Good	Good	Good
knowledge-based river generalization	Ordinary	Ordinary	Good	Good
River networkgeneralization by catchment area	Good	Good	Good	Good
River selection by the BP neural network	Good	Ordinary	Good	Good
River selection by pattern recognition	Good	Good	Good	Good

abided by in river network generalization. Last, it presents six approaches for river network generalization, i.e. the approach for select river segments by indices, the approach for select river segments by the river tree, a knowledge-based approach to river network generalization, an approach to generalizing river networks by catchment area, an approach to river selection by the BP neural network algorithm, an algorithm for river network selection based on structure and pattern recognition, respectively.

References

Ai T., Liu Y., Chen J., 2006, The hierarchical watershed partitioning and data simplification of river network, In Progress in Spatial Data Handling, Riedl A., Kainz W., Elmes G.A., Eds.; Springer: Berlin/Heidelberg, Germany, pp: 617–632.

Ai T., Liu Y., Huang Y., 2007, The hierarchical watershed partitioning and generalization of river network, Acta Geodaetica Et CartographicaSinica, 36(2): 231–236.

Bard S., Ruas A., 2005, Why and how evaluating generalised data? In Developments in Spatial Data Handling; Springer: Berlin/Heidelberg, Germany, pp: 327–342.

Buttenfield B.P., Stanislawski L.V., Brewer C.A., 2010, Multiscale representations of water: tailoring generalization sequences to specific physiographic regimes. In Proceedings of the GIScience 2010, Zurich, Switzerland, 15–17 September 2010, pp: 14–17.

Charlton R., 2007, Fundamentals of Fluvial Geomorphology; Routledge: London, UK.

Chen Y., Wilson J.P., Zhu Q., Zhou Q., 2012, Comparison of drainage-constrained methods for DEM generalization, Computers and Geosciences, 48: 41–49.

De Serres B.,Roy A.G., 1990, Flow direction and branchinggeometry at junctions in dendritic river networks, The Professional Geographer, 42(2):149–201.

Ethem A., 2010, Introduction to Machine Learning, MIT Press.

Feder J., 1988, Fractals. Plenum Press, New York.

Fiorentino M. and Claps P., 1992, On what can be explained by the entropy of a channel network. In: V.P. Singh and M. Fiorentino (Editors). Entropy and Energy Dissipation in Water Resources. Kluwer, Dordrecht, The Netherlands, pp. 139–154.

Fiorentino M., Claps P. and Singh V.P., 1993, An entropy-based morphological analysis of river-basin networks. Water Resources Research, 29(4): 1215–1224.

Génevaux J.D., Galin É., Guérin E., Peytavie A., Beneš B., 2013, Terrain generation using procedural models based on hydrology. ACM Trans. Graph. 32: 1–10.

Hack J., 1957, Studies of longitudinal stream profiles in Virginia and Maryland, U.S. Geological Survey Professional Paper, 1957, 294-B.

He Z., 2004, Principles and methods of Map data processing models, Wuhan: Wuhan University Process. (in Chinese)

Hentschel H.G.E., Procaccia I., 1983, The infinite number of dimensions of fractals and strange attractors, Physica, 8D: 435–444.

Horton R.E., 1945, Erosional development of streams and their drainage basins: hydro-physical approach to quantitative morphology, Geological Society of America Bulletin, 56 (3): 275–370.

LaBarbera P., Rosso R., 1987, The fractal geometry of river networks, Eos Transactions on AGU, 68(44): 12–76.

Li Z., Zhu Q., Gold C., 2010, Digital Terrain Modeling: Principles and Methodology; CRC Press: Boca Raton, Florida, USA.

Lovejoy S., Schertzer D., Tsonis A.A., 1987, Functional box counting and multiple elliptical dimensions in rain, Science, 235: 1036–1038.

Mandelbrot B.B., 1983, The Fractal Geometry of Nature, W. H. Freeman, New York.

Mandelbrot B.B., 1986, Self-affine fractal sets. in: L. Pietronero and E. Tosatti (Editors). Fractals in Physics. North-Holland, Amsterdam, pp. 3–28.

Ritter M.E., 2016, The physical environment: an introduction to physical geography, available online: http://www.earthonlinemedia.com/ebooks/tpe_3e/title_page.html (accessible on 28 November 2017).

Scheidegger A.E., 1967, On the topology of river nets, Water Resources Research, 3(1): 103–106.

Shannon C.E., 1948, The mathematical theory of communications, I and II. Bell System Technical Journal, 27: 379–423.

Shao L., He Z., Ai Z., Song X., 2004, Automatic generalization of river network based on BP neural network techniques, Geomatics and Information Science of Wuhan University, 29(6): 555–557.

Shreve R., 1966, Statistical law of stream numbers, Journal of Geology, 74 (1) :17–37.

Shreve, R. L., 1967, Infinite topologically random channel networks, Journal of Geology, 75:178–186.

Stanislawski L.V., 2009, Feature pruning by upstream drainage area to support automated generalization of the United States National Hydrography Dataset. Computers, Environment and Urban Systems, 33: 325–333.

Strahler A.N., 1952, Hypsometric (area altitude) analysis of erosional topography, Geological Society of America Bulletin, 63: 1117–1142.

Strahler A.N., 1957, Quantitative analysis of watershed geomorphology, Transactions of the American Geophysical Union, 38 (6): 913–920.

Tarboton D.G., Bras R.L., Rodriguez-Iturbe I., 1998, The fractal of nature of river networks, Water Resources Research, 24(8): 1317–1322,

Thomson R., Brooks R., 2000, Efficient generalization and abstraction of network data using perceptual grouping. In the proceedings of the 5th GeoComputation. University of Greenwich, Kent U.K.

Töpfer, F., Pillewizer, W., 1966. The principles of selection. The Cartographic Journal 3 (1), 10–16.

Touya G. 2006.: A Method for generalization of river networks based on "strokes" and database enrichment.Extended Abstracts Proceedings of 4th International Conference GIScience 2006, Münster, Germany. pp 191–194.

Twidale C.R., 2004, River patterns and their meaning, Earth Science Review, 67: 159–218.

Wu F., Tan X., Zhan R., Wang L., 2007, Automated selection of river networks based on knowledge representation and inferences, Journal of Liaoning University of Engineering and Technology, 26(2): 183–186.
Zhang Q.N., Quan H., 2005, Construction and application of river tree, ActaScientiarumNaturalium UniversitatisSunyatseni, 44(6): 101–104.
Zhang Q.N., 2006, Generalization of drainage network with density differences, Acta Geodaetica et Cartographic Sinica, 35(2): 191–196.

Chapter 6
Description and Generalization of Continuous Areal Features

6.1 Introduction to Continuous Areal Features

Areal features on maps are visual representation of real phenomenon in the geographic space. Areal phenomena are two dimensional in spatial extent, having both length and width and appearing on maps as polygons (Yan et al. 2008). An example of areal phenomena is a lake, assuming that only its two-dimensional surface is concerned. Political units, such as countries, provinces/states and counties, can also fit this framework if they are represented on large or intermediate scale maps, because each of them is symbolized as an enclosed region. However, an insight into the differences between lakes and political regions on maps may gains that their distribution modes are different: lakes are shown as discrete polygons, while political regions are usually visualized using continuous polygons (Fig. 6.1).

Discrete areal features are well defined and are easy to locate, measure, and count, and their edges or boundaries are readily defined. A single discrete areal feature is usually modelled using a polygon consisting of a number of vertices whose coordinates can be saved spatial databases. Examples of discrete areal features on large or intermediate scale maps include lakes, ponds, seas, rivers, buildings, parks, squares, playgrounds etc.

Continuous areal features, on the other hand, are more difficult to locate and less well defined because they exist across space. The most commonly cited examples of continuous areal features are temperature and elevation, because changes in both temperature and elevation tend to be gradual over relatively large areas, and therefore it is difficult to define their locations and distributions. Land cover is another example of continuous areal features which is easier to locate, represent and define. As shown in Fig. 6.2, land patches are represented using polygons filled with different colors denoting different types of land. Borders of land patches are easy to recognize and the vertices of border polygons can be collected by surveying techniques and saved in a database. Similarly, the other continuous phenomena (e.g. provincially average income, precipitation and air pressure) can also be

H. Yan, *Description Approaches and Automated Generalization Algorithms for Groups of Map Objects*, https://doi.org/10.1007/978-981-13-3678-2_6

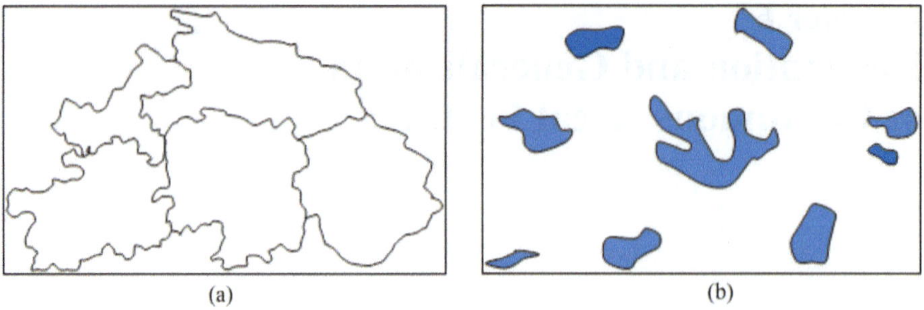

Fig. 6.1 Areal features on maps have two types of distribution modes: continuous and discrete. (**a**) Political regions on the map are shown as continuous polygons. (**b**) Lakes on the map are shown as discrete polygons

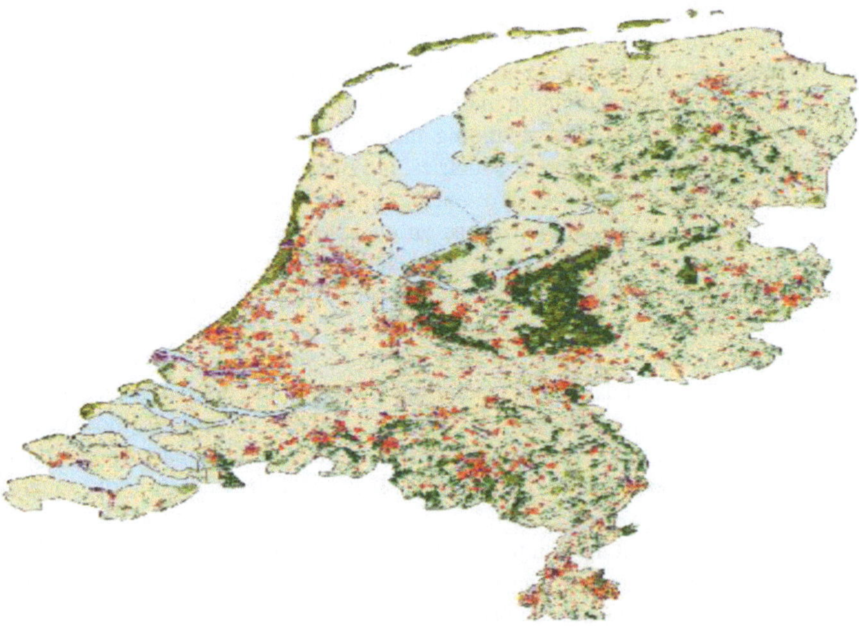

Fig. 6.2 A land use map consisting of continuous polygons, different filling colors representing different types of land

represented using continuous polygons. For example, in Fig. 6.3, the topography is visualized by a hypsometric map which mainly consisting colored polygons. In this map, as far as a polygon is concerned, the border of the polygon is an enclosed contour line which actually forms a polygon, and the color filled in the polygon means that the elevations of this polygon are approximately the same.

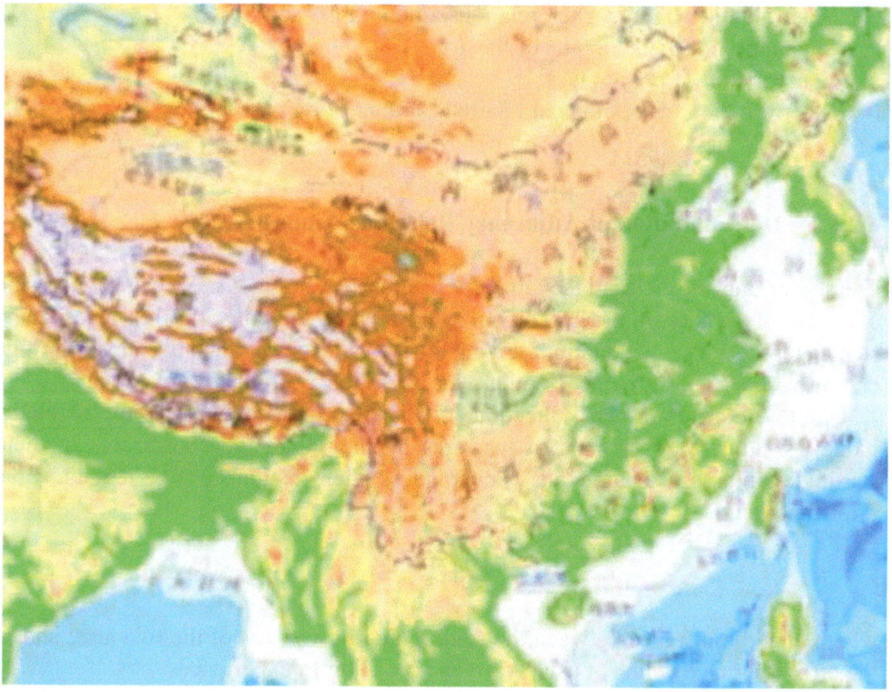

Fig. 6.3 Topography can be represented using continuous polygons filled with colors representing different elevations

This focuses on description and generalization of continuous areal features on maps, mainly discussing categorical maps.

6.2 Description of Continuous Areal Features

6.2.1 Topological Relations Among Continuous Polygons

Topological relation is a character owned by spatial but not non-spatial data. Construction of topological relation is not only useful in map generalization but also necessary in establishing spatial databases. Existing achievements have revealed that construction of topological polygons using arcs is one of the most difficult problems in topological relations. Topological polygons can be built either manually (e.g. polygons are manually built in the USA's DIME system) or automatically (i.e. polygons are built by software). This chapter has no interest in manual methods but pays attention to how to automatically build polygons using arcs. The following section presents an algorithm proposed by Yan et al. (2000). This algorithm is based on azimuths and included angles of arcs. It consists of four steps:

(1) Determination of adjacency relations among arcs
(2) Calculation of the azimuths of the arcs
(3) Construction of polygons
(4) Determination of topological relations among polygons

6.2.1.1 Determination of Adjacency Relations Among Arcs

The source data are arcs and they may form connected polygons. Each arc consists of a number of vertices. In order to organize the arcs to form polygons, it is a necessary step to get the adjacency relations among the arcs, i.e. to get which two arcs' starting/ending vertices are the same. In other words, if their coordinates are equal.

Due to the errors in data collection such as map digitization, it is impossible to judge the adjacency relations of two arcs by judging if the coordinates of their starting/ending vertices are equal. Nevertheless, their relations can be judged by an error function.

$$d \leq t$$

where, d is the distance between the starting/ending vertices of the two arcs; and is the error tolerance which is usually given according to experiences.

Here, d is calculated by the coordinates of the starting/ending vertices of the two arcs.

For example, in Fig. 6.4, the distance between V_1 and V_2 is less than the tolerance value t which is equal to the diameter of the circle in Fig. 6.4, so the two arcs topologically adjacent at V_1 and V_2.

All the topological adjacency relations can be got in this way, by which all arcs may be connected.

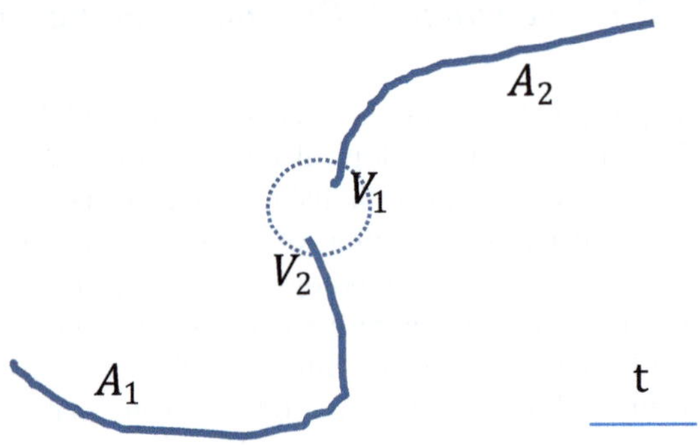

Fig. 6.4 Judging adjacency relations between two ending vertices of two arcs by an error function

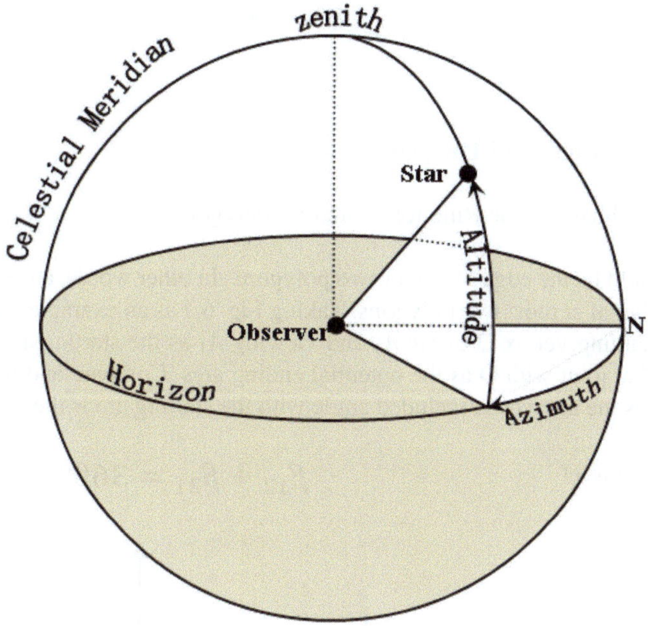

Fig. 6.5 Definition of azimuth

6.2.1.2 Calculation of the Azimuths of the Arcs

Traditionally, an azimuth is defined as an angular measurement in a spherical coordinate system. The vector from an observer (i.e. the origin) to the point of interest is projected perpendicularly onto a plan. The angle between the projected vector and a reference vector on the reference plane is called the azimuth. An example of azimuth is the angular direction of a star in the sky. The star is the point of interest, the reference plane is the local horizontal area, and the reference vector points north. The azimuth is the angle between the north vector and the star's vector on the horizontal plane (See Fig. 6.5).

In surveying techniques, the concept of azimuth can be defined in a Cartesian Plan Coordinate System. the azimuth of a radius A_1 (α_1 in Fig. 6.6) is defined as the angle rotating from the positive half of the horizontal axis to A_1 in counter clockwise (so it is α_1 in Fig. 6.6). The value of an azimuth falls in $[0^{\circ}, 360^{\circ}]$.

The included angle β_{21}, the angle rotating from A_1 to A_2 in counter clockwise, can be calculated by:

$$\beta_{21} = \begin{cases} \alpha_2 - \alpha_1, \text{if } \alpha_2 \geq \alpha_1 \\ \alpha_2 - \alpha_1 + 360^o \end{cases} \tag{6.1}$$

It is apparent that

$$\beta_{12} + \beta_{21} = 360^o \qquad (6.2)$$

6.2.1.3 Construction of Polygons

Minimum Included Angle Rule for Searching Polygons

One arc should be the edge of one or two polygons. In other words, an arc can be the common edge of at most two polygons. Taking Fig. 6.7 as an example, starting from one of the ending vertex O of arc A_1 and viewing A_1 as the starting arc, regard the other arcs that joint with O as the potential ending arcs. Compare and find from the potential arcs the arc whose included angle with the starting arc is the minimum and

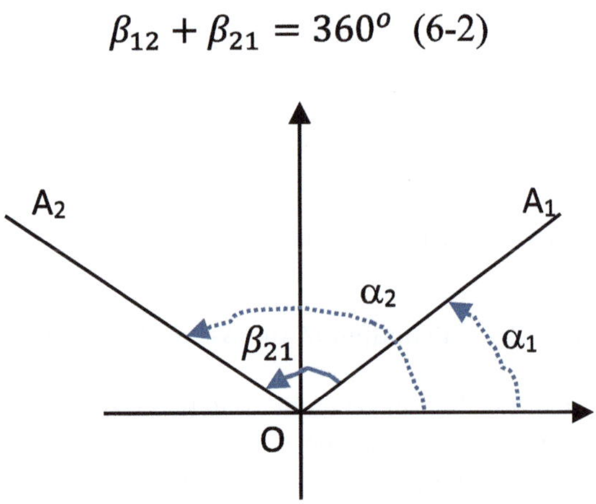

Fig. 6.6 Definition of included angle

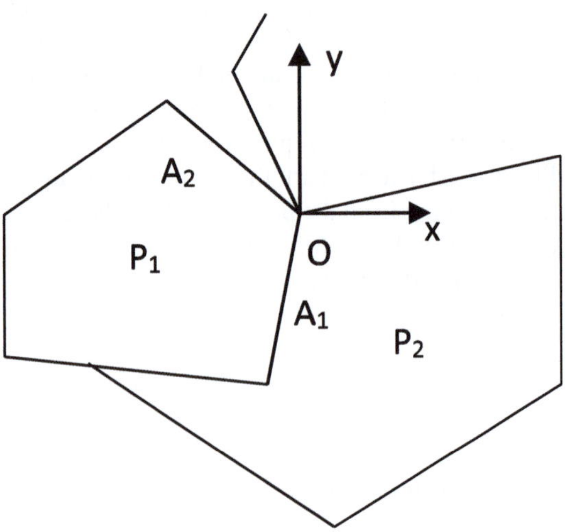

Fig. 6.7 The minimum included angle rule for searching polygons

Table 6.1 Adjacency relations

Arc number	The arcs adjacent to its starting vertex	The arcs adjacent to its ending vertex
1	4, 8	−2, −7
2	−3, 6	−1, −7
3	−4, 5	2, 6
4	1, 8	5, 3

Fig. 6.8 Adjacent arcs

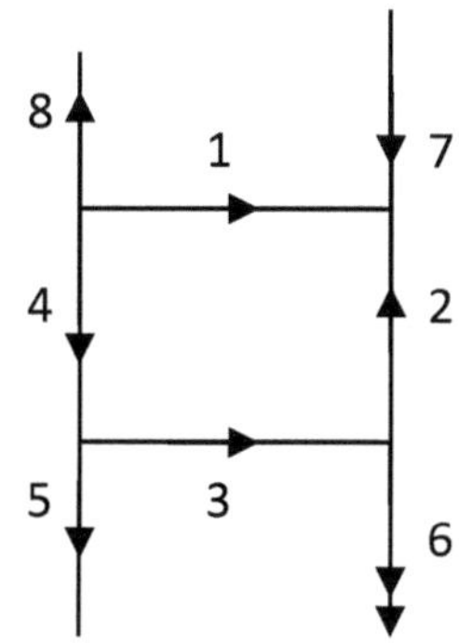

view this arc (i.e. A_2 in this example) as the successive arc in construction the polygon. This process for getting the successive arc is called "Minimum included angle rule for searching polygons".

Taking A_2 as the new starting arc and search for its successive arc abiding by the minimum included angle rule for searching polygons, until the new arc is A_1. The arc chain got in this searching process consist a polygon.

Similarly, start from O and take A_1 as the new ending arc and take the other arcs joint with O as the potential ending arcs. Compare and find an arc from the potential arcs whose included angle with the ending arc is the minimum, then take this arc as the new ending arc and repeat this searching process, until the new ending arc is A_1.

Algorithm for Constructing Polygons

Step 1: establishment of the adjacency relation among arcs

The objective of this step is to find the adjacency relations among arcs and save them in 2-dimensional tables. Such an example is shown in Table 6.1 which representing the adjacency relations among arcs in Fig. 6.8.

It should be noticed that in Table 6.1 if the starting vertex of an arc is joint with another arc, its arc number is recorded as positive; otherwise, its arc number is recorded as negative.

Step 2: calculation of azimuths

Table 6.2 Joint relation among arcs and polygons obtained by the minimum included angle rule for searching polygons

No. of polygon	The number of the arcs/edges consisting of the polygon	Arcs'/edges' numbers
1	7	−1,4, −15, −14, −9, −8, −6
2	3	−1, 3, 18, 11
3	1	−2
4	1	−2
5	3	−3, 4, 5
6	3	−3, 1, −11, −18
7	7	−4, 1, 6, 8, 9, 14, 15
8	3	−4, 3, −5
9	4	−5, −15, −13, −12, −18
10	3	−5, −4, 3
11	7	−6, −1, 4, −15, −14, −9, −8
……	……	……
30	7	−15, −14, −9, −8, −6, −1,4
31	1	−16
32	1	−16
33	1	−17
34	1	−17

The azimuths of all starting and ending line segment of the arcs are calculated and saved in a table. This may facilitate the calculation of the included angles of the arcs.

Step 3: connecting arcs to form polygons

This step is done abiding by the minimum included angle rule for searching polygons.

Starting from the starting vertex of the arc numbered 1, search for the adjacency relations among arcs (see Table 6.1) and find the arcs joint to the starting vertex of the arc numbered 1, and form two polygons (form only one polygon if this starting vertex is only joint to one arc).

Take another arc as the starting arc and repeat this process, and totally get 2N polygons (N is the arc number). In the arc searching process, the arc numbers that form each polygon should be recorded and saved in a 2-dimensional array (say, A). It should be notice that if an arc's starting vertex is joint with its previous arc this arc's number is recorded as positive; otherwise, it is recorded as negative (see Table 6.2).

Step 4: deletion of redundant and incorrect polygons

Some of the polygons generated by the minimum included angles for searching polygons are repetitive (e.g. an island polygon is searched and recorded for two

Table 6.3 Resulting polygons of Table 6.2

No. of polygon	The number of the arcs consisting of the polygon	Arcs' numbers
1	4	−1, 3, 18, 11
2	1	−2
3	3	−3, 4, 5
4	4	−5, −15, −13, −12, −18
5	6	−6, −11, −12, 10, −9, 7
6	2	−7, −8
7	3	−10, 13, −14
8	1	−16
9	1	−17

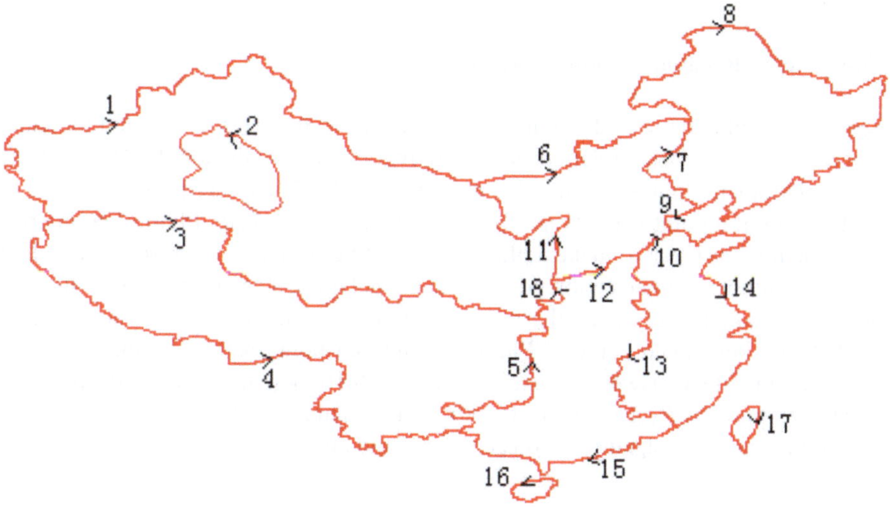

Fig. 6.9 Polygons consisting of directed arcs

times), and some are incorrect (e.g. the border polygon). These repetitive and error polygons should be deleted from the array and only the true polygons should be left.

To delete the redundant polygons, the array that records arcs forming polygons are search and compared. If any two polygons have the equal number of arcs and their corresponding arcs' numbers' absolute values are also equal, delete arbitrary one of the two polygons from. The incorrect polygons can be found by this way: if a polygon has a common edge (i.e. an arc) with another polygon and their have topological containment relations, delete the bigger polygon from the array.

After the redundant and incorrect polygons in Table 6.2 have been deleted, we can get Table 6.3.

The arcs used in this process are directed ones, and therefore the arcs can be viewed as a directed graph (See Fig. 6.9 for an example).

6.2.1.4 Determination of Topological Relations Among Polygons

After determination of the topological relations among the arcs and the polygons (Table 6.2 and Table 6.3), construction of the polygons using the arcs has been completed. This step determines the relations among polygons themselves.

Neighbouring Relations Among Polygons

Search the table that record the arc-polygon relations (e.g. Table 6.3) and record the neighbouring relations if any two polygons have a common edge/arc.

Containment Relations Among Polygons

Search the table that record the arc-polygon relations (e.g. Table 6.3) and the table that record neighbouring relations among polygons. Record the topological containment relations if a polygon's interior point is included in another polygon.

To judge if a point is contained by a polygon, the so-called "Plumb line interior point method" is used: to make a plumb line starting from a point, if this plumb line has even number of intersection points with the polygon, it can be concluded that the point is outside the polygon; otherwise, the point is an interior point of the polygon.

It should be noticed that the "Plumb line interior point method" sometimes may lead to an incorrect answer. For example, in Fig. 6.10, polygon P_1 contains polygon P_2 and P_2 contains P_3. However, it is not correct that P_1 contains P_3.

By this way, the island polygons can be found and recorded.

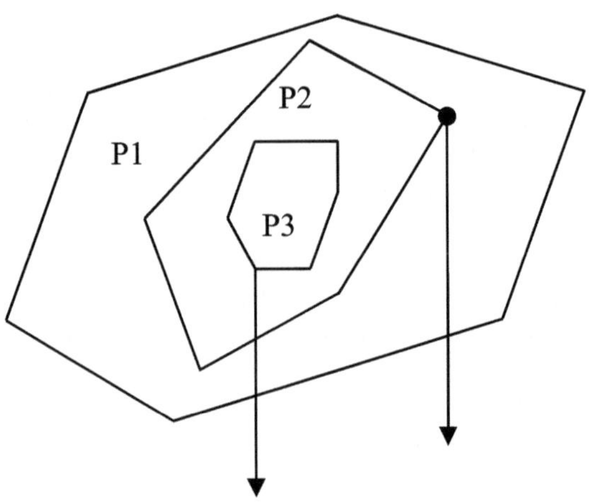

Fig. 6.10 Containment relations among polygons

6.2.2 Topological Relations on Land-Use Maps

Areal patches on land-use maps which belong to categorical maps have four characters in spatial distribution: firstly, the areal patches are continuous polygons on the map; secondly, the mapping areas of the map are all covered by polygonal land patches; thirdly, it is common that polygonal land patches have multi-layer of containment relations; lastly, semantic relations among land patches are hierarchical (Bader and Weibel 1997).

According to the four characters, topological relations among areal land patches can be classified into four categories: separated, neighbouring, containing and contained.

To record the topological relations, the following data structure for describing land patches can be defined and used in land-use data files, in C++:

```
typedef struct tagPatchDescription //PatchDescription is the name of
the structure
// for describing a land patch
{
  int ID;//ID of the land patch
  char TypeCode[4];//code of the type of the land patch
  char Name[30];//name of the land patch
  char Layer [4];//code of the layer of the land patch
  int Num;//the number of vertices on the edges of the land patch polygon
  CDou3DPoint pAllCDou3DPoint;// vertices on the edges of the land patch
polygon
  RECT MBR; //minimum bounding rectangle of the land patch polygon.
  int AdjPatchNum; // the number of the land patches adjacent to this land
patch
  LONGARRAY *AdjPatchID;//IDs of the land patches adjacent to this land
patch
  LONGARRAY *IncludedAdjPatchID;//IDs of the land patch included by this
land patch
  LONGARRAY *BeIncludedAdjPatchID;//ID of the land patch that includes
this patch
}PatchDescription
```

6.2.3 Neighbouring Relations on Land-Use Maps

Ai and Liu (2002) proposed that merging of neighbouring land patches is one of the basic rules in land-use data generalization; however, how to judge neighbouring relations among polygonal land patches is a difficult problem. Neighbouring relations are at least related to topological, geometric, semantic and Gestalt factors in

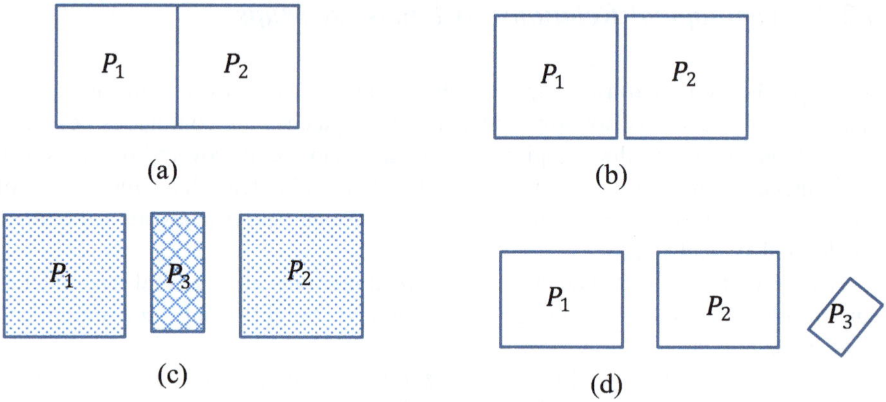

Fig. 6.11 Four types of neighbouring relations: (**a**) $P_1$and P_2 have topological neighbouring relations, because they have a common edge; (**b**) $P_1$and P_2 have geometric neighbouring relations, because their distance is small; (**c**) $P_1$and P_2 have semantic neighbouring relations, because they belong to the same type of land; and (**d**) $P_1$and P_2 have Gestalt neighbouring relations, because they are similar in direction, size and shape compared with P_3

spatial cognition, and correspondingly there are four types of neighbouring relations (Yan et al. 2008):

- Topological neighbouring relation: two polygons that own at least a common edge are topologically neighbouring (Fig. 6.11a).
- Geometrical neighbouring relation: two polygons (though they are not topologically neighbouring) adjacent in distance are called geometrically neighbouring (Fig. 6.11b).
- Semantic neighbouring relation: two polygonal land patches belonging to same or similar types of lands are called semantically neighbouring (Fig. 6.11c).
- Gestalt neighbouring relation: if two polygonal objects are similar in direction, size and/or shape, they are called Gestalt neighbours (Fig. 6.11d).

It is difficult to quantitatively order the neighbouring relations of one areal land with its neighbours. Fortunately, in most cases it is unnecessary to quantitatively measure the neighbouring relation between two areal land patches; on the contrary, a qualitative description, i.e. neighbouring or non-neighbouring, is enough in land patches generalization.

To judge if other land patches are neighbouring with a specified land patch (say, A), the buffering method is often used: firstly, the polygons that are topologically neighbouring with A are neighbours of A; secondly, make a buffer of A using a given distance (i.e. the threshold). If an areal land patch falls in the buffering polygon, it is regarded as a neighbouring land patch of A. For example, in Fig. 6.12, A is topologically neighbouring with B, C, D, E and F; and it is a neighghour of G, for G falls in A's buffering polygon (while A is not a neighbour of H, for its buffering polygon does not cover any part of H).

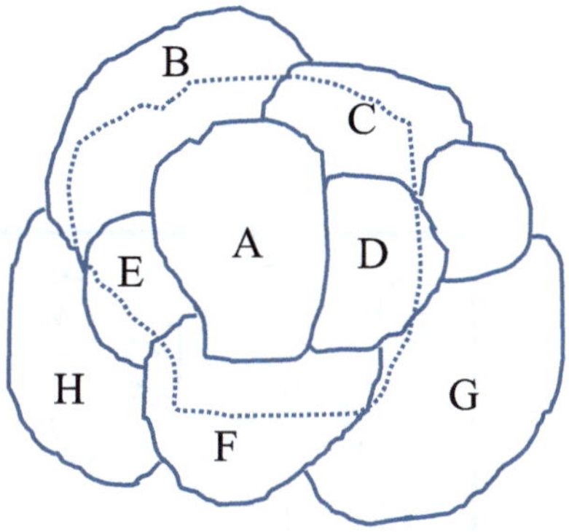

Fig. 6.12 Judging neighbouring relations by the buffering method

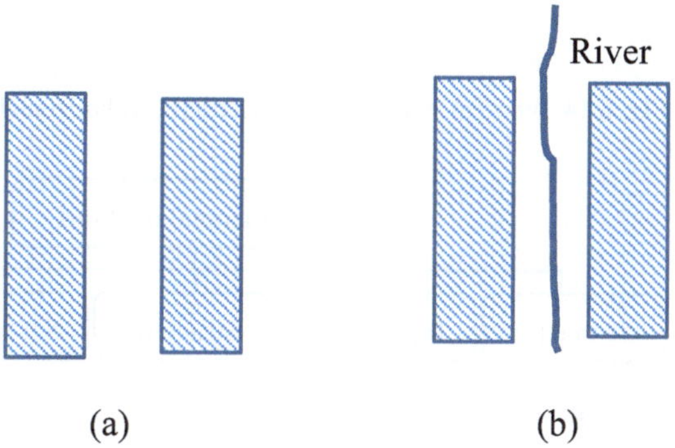

Fig. 6.13 Context is a factor in determining neighbouring relations: (**a**) two settlements are neighbouring with each other; and (**b**) when a river passes through the two settlements, they are no long neighbouring with each other, though their distance does not change

Nevertheless, neighbouring relation is not only a simple concept defined by distance; contexts are also unneglectable in many cases. For example, if only by the given distance threshold, A is neighbouring with B. However, if another object C appears between A and B, A and B are no longer neighbours (Fig. 6.13).

As far as a group of areal land patches (i.e. a set of areal land patches, say, $\{P_1, P_2 \ldots\ldots, P_n\}$) are concerned, P_1 and P_2 are regarded as neighbours if they meet one of the following criterions:

1. P_1 and P_2 are topological neighbours; and
2. P_1 and P_2 have the same semantic attributes, and their distance is less than a given threshold, and their Voronoi polygons are neighbours, though they are topologically separated themselves.

6.2.4 *Horizontal and Vertical Relations on Land-Use Maps*

Neun and Steiniger (2005) proposed that there are two types of relations on land-use maps: vertical relations and horizontal relations. The former is defined for multi-scale maps, and it exists among the features on maps at different scales; the latter

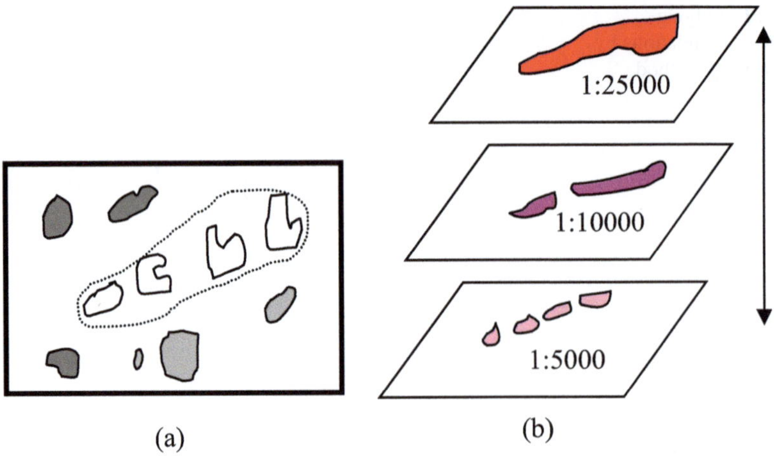

Fig. 6.14 Vertical and horizontal relations: (**a**) horizontal relations; and (**b**) vertical relations

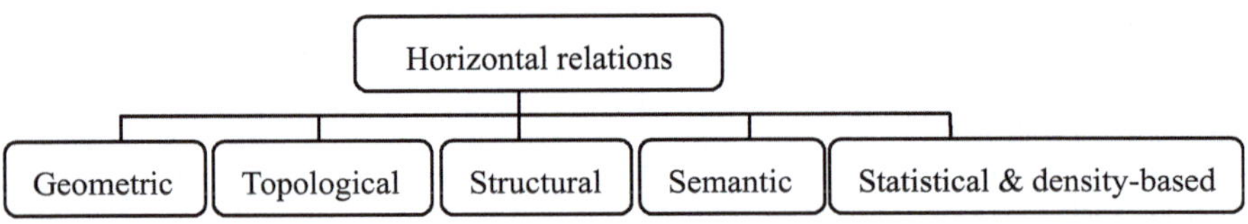

Fig. 6.15 Horizontal relations

exists among different features on the same maps at the same scale or different features in the same map library (Fig. 6.14).

Horizontal relations on the map may be geometric, topological, structural, semantic, statistical and density-based (Fig. 6.15).

- Geometric relations are used to describe the sizes, locations, shapes and directions among objects. The factors that are usually used include area, perimeter, number of edges, number of vertices, length, deviation angle etc.
- Topological relations include topological adjacency/neighbour, topological intersection (e.g. overlap, touch, pass-through), topological containment, topological separation etc.
- Structural relations are closely related to human's visual sense and generally guided by the Gestalt's principles. Theses relations include foreground-background relations, micro-scale structural relations, meso-scale structural relations, macro-scale structural relations etc.
- Semantic relations include semantic similarity relations, semantic priority relations, semantic attraction/repellence relations, semantic logic relations etc.
- Statistical and density-based relations mainly include a number of factors from landscape ecology and statistics, e.g. mean area, mean length, average number of vertices/edges of a polygon, type of the land patch, etc.

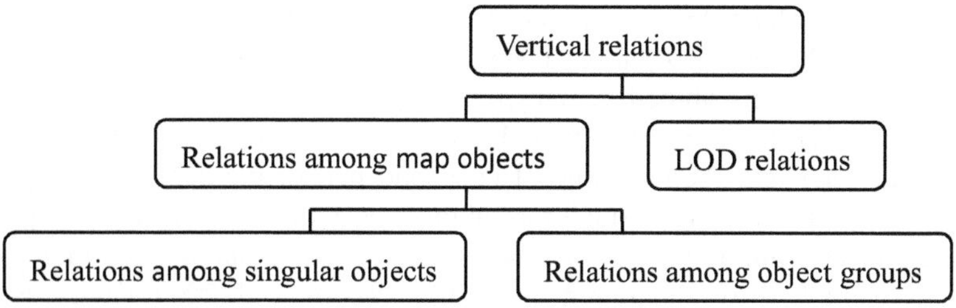

Fig. 6.16 Vertical relations

Vertical relations include LOD (level of detail) relations and the relations among map objects (Fig. 6.16). LOD relations are the relations among map object in the process of map scale change which can be usually described using a topological neighbouring matrix for expressing the change of the metadata of the whole map, a matrix for expressing the hierarchical relations among the types of map features. The relations among map objects include the ones among singular objects and the one among object groups or object clusters in the process of or after map scale change.

6.2.5 *Geospatial Semantic Relations on Land-Use/ Cover Maps*

6.2.5.1 Fundamental Theories in Geospatial Semantics

The term semantics refers to the meaning of things and is in contrast with the term syntactics (Hu 2017). Geospatial semantics adds the adjective "geospatial" in front of "semantics", and this addition both restricts and extends the initial applicable area of semantics. On one hand, geospatial semantics manifest that it focuses on the expressions that have a close connection with geography rather than any other general expressions; on the other hand, geospatial semantics pays its attentions to not only linguistic expressions but also the meaning of geographic places, geospatial data, and the GeoWeb. In this sense, studies on geospatial semantics usually care about how to understand the meaning of geographic entities as well as their counterparts in the cognitive and digital world, such as cognitive geographic concepts and digital gazetteers (Ahlqvist et al. 2017).

Geospatial semantics has been a recognized subfield in GIScience for years (Agarwal 2005; Mark et al. 2000) which involves a variety of research areas. Kuhn (2005) defines geospatial semantics as "understanding GIS contents, and capturing this understanding in formal theories." Obviously, this definition may be divided into two parts: understanding and formalization. The understanding here triggers an interesting question: who is supposed to understand the GIS contents, people or machines? If our answer is "people", geospatial semantics is apparently connected to human cognition of geographic concepts and spatial relations

(Egenhofer and Mark 1995; Smith and Mark 2001; Golledge 2002); whereas if our answer is "machines", geospatial semantics may involve the study on the semantic interoperability of distributed systems, digital gazetteers, and geographic information retrieval (Bishr 1998; Harvey et al. 1999; Fonseca et al. 2002; Goodchild and Hill 2008; Jones and Purves 2008). The second part of the definition proposes to capture this understanding through formal theories. Ontologies, as formal specifications of concepts and relations, have been widely studied and applied in geospatial semantics (Frank 2001; Pundt and Bishr 2002; Visser et al. 2002; Couclelis 2010) and formal logics, such as first-order logic and description logics (Hitzler et al. 2009), are often employed to define the concepts and axioms in an ontology. While Kuhn's definition includes these two parts, research in geospatial semantics is not required to have both: one focuses on understanding, while another examines formalization.

Research on geospatial semantic has absorbed a lot of nutrients from the Advances in computer and information technologies, especially the Web. With the Semantic Web initially proposed by Berners-Lee et al. (2001) and Egenhofer (2002) envisioned the Geospatial Semantic Web that can understand the semantics of geospatial requests from users and automatically obtain relevant results. The development of Linked Data (Bizer et al. 2009) as well as the resulting Linked Open Data cloud (Heath and Bizer 2011) have fostered geospatial semantic studies on organizing, publishing, retrieving, and reusing geospatial data as structured Linked Data (Janowicz et al. 2013). In the meanwhile, there is a rapid increase in the volume of unstructured natural language texts on the Web, such as social media posts, blogs, and Wikipedia entries. While often subjective, textual data reveal the understanding and perceptions of people towards natural and social environments. Existing studies have demonstrated the use of unstructured text data in extracting place semantics and understanding the spatiotemporal interaction patterns between people and places (Adams and McKenzie 2013; Ballatore and Adams 2015; Hu et al. 2015). More novel research topics based on big text data have become possible which causes the fast development of natural language processing and text mining techniques.

Geospatial semantics is a broad field that adopts a unique research perspective towards geospatial problems. To some extent, geospatial semantics can be compared with geospatial statistics: both can be applied to various problems across domains and both have their own unique set of methods (e.g., ontological modeling and natural language processing for geospatial semantics). In recent years, a lot of research on geospatial semantics has been conducted, and the results are published in journals or presented in conferences, such as the Conference on Spatial Information Theory (COSIT), the International Conference on Geographic Information Science (GIScience), the International Conference on Geospatial Semantics (GeoS), and many others. To sum up, an overview of existing achievements tells that the following research areas in geospatial semantics worth our noticing.

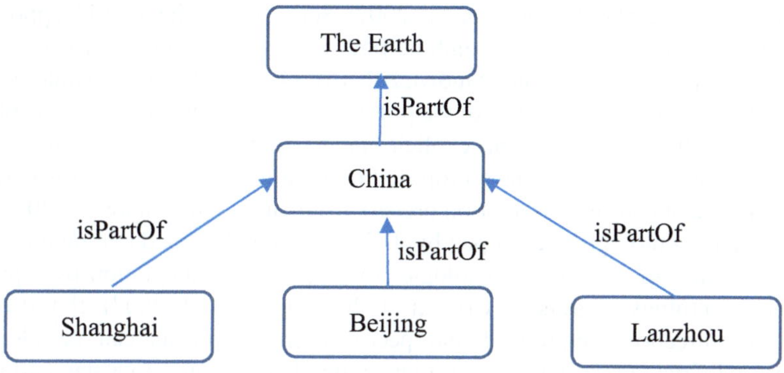

Fig. 6.17 A fragment of an ontology

Semantic Interoperability and Ontologies

Semantic interoperability is driven by the applications of GIS. Traditional GISs were used locally and their functions and data were increasingly encapsulated into services and shared via the Web. As a result, it becomes necessary to formalize the semantics of the distributed services so that they can automatically interact with each other and be dynamically integrated. In addition, semantic interoperability is also critical for spatial data infrastructures of states, nations, cities or/and counties that provide access to a wealth of distributed geospatial data sources and services which can be combined for various queries and tasks (Alameh 2003; Lemmens et al. 2006; Lutz et al. 2009).

A major approach to enable geospatial semantics is to develop ontologies. Ontologies in geospatial semantics are similar to those in computer science and bioinformatics, which serve the function of formalizing the meaning of concepts in a machine-understandable manner (Couclelis 2009). From a data structure perspective, an ontology can be considered a graph with concepts as nodes and relations as edges (Fig. 6.17).

Ontologies may be embedded into corresponding geographic information systems and/or Web services as an additional component to enable semantic interoperability (Frank 1997). Ontologies should be developed before they are embedded into geographic information systems. Three types of ontologies can be identified from the literature: top-level ontology, domain ontology, and ontology design pattern (Ashburner et al. 2000). The process of developing ontologies is called ontology engineering. Three types of approaches are often used in ontology engineering: top-down, bottom-up, and hybrid approaches. Top-down approaches rely on knowledge engineers and domain experts to define and formalize the ontological concepts and relations (Brodaric 2004; Schuurman and Leszczynski 2006; Gates et al. 2007; Shankar et al. 2007; Wang et al. 2007). Bottom-up approaches employ data mining methods to extract concepts and relations from structured databases or unstructured natural language texts (Maedche and Staab 2004; Shamsfard and Barforoush 2004;

Buitelaar et al. 2005; Baglioni et al. 2007; Sen 2007). The hybrid approaches integrate the previous two and combine both expert knowledge and results from data mining processes (Hu and Janowicz 2016). One challenge in ontology engineering is to define the semantics of the primitive terms (i.e. the atomic concepts that cannot be further divided) in an explicit and unambiguous way. To address this challenge, some researchers have proposed to ground the primitive terms based on the environment and the observation process (Mallenby 2007; Janowicz 2012).

With many ontologies developed by different researchers and communities, it is often necessary to align these ontologies to support data integration by a process known as ontology alignment (Cruz et al. 2004; Hess et al. 2007). Based on the alignment direction, centralized and peer-to-peer alignments can be identified (Sunna and Cruz 2007). The former aligns multiple ontologies to a standard ontology, while the latter establishes links between two peer ontologies. The alignment methods can be classified into: element-level, structure-level, and hybrid methods (Shvaiko and Euzenat 2005).

Although a lot of achievements have been made on ontologies, it is worth noting that developing ontologies is only one approach for realizing semantic interoperability. In fact, some researchers have criticized the use of ontologies to address semantic issues by arguing that ontologies as priori agreements cannot capture the meaning of concepts that change dynamically (Di Donato 2010). Thus, new approaches for semantic interoperability may also be possible and need further investigation.

Digital Gazetteers

Digital gazetteers are structured dictionaries for named places. The building of digital gazetteers, that organize details of named places, is a growing area of research that necessarily involves cross-domain issues from fields such as spatial cognition, geographic information science, social history, computer science, and geographic information retrieval. Gazetteer-based services or applications have been and are being developed and deployed in many specific fields such as public health, natural history data management, cultural history, and automated georeferencing of texts (geoparsing). Some of these gazetteers have developed into sources of authoritative names and have been used widely, and the organizations that have prepared them, mostly for governmental purposes, have been and/or are being known as toponymic authorities.

The place entries within a digital gazetteer are often organized into a graph, with nodes representing places and edges capturing their relations (e.g., Beijing is a part of China in Fig. 6.17). In fact, digital gazetteers can be considered as a special type of ontology. As people frequently use place names rather than numeric coordinates to refer to places, digital gazetteers fill the critical gap between informal human discourses and formal geographic representations. From a perspective of geospatial semantics, digital gazetteers help machines understand the geographic meaning (e.g., the spatial footprint) of a textual place name as well as the relations among places.

One important topic in gazetteer research is enriching existing gazetteers with local or vernacular place entries, which refers to the addition of local place names used in everyday conversations into the gazetteer, for such vernacular places are important for some GIS applications. Since these places often do not have clearly defined boundaries, research has been conducted to represent their vague spatial footprints.

Another topic in digital gazetteer research is to align and conflate multiple gazetteers. Digital gazetteers from independent sources may have different geographic coverages, different spatial footprints (e.g., the same place may be represented as a point or as a polygon in different gazetteers), different place types, and different attributes. While these differences can be combined to form a richer data resource, they also present challenges for gazetteer conflation. Based on the conflating targets, their schema-level and instance-level conflations can be identified. The schema-level conflation aligns the place types from one gazetteer to that of another gazetteer and can be considered as a special type of ontology alignment. Naturally, the ontology alignment methods, such as those based on the similarities of labels, definitions, and structures, can be employed to align the place types in different gazetteers. There are also methods that leverage the spatial distribution patterns of place instances belonging to a place type to align place types. The instance-level conflation aims at merging the specific place entries in different gazetteers. There exist a variety of methods for measuring the similarities of spatial footprints (geometries), place types (ontologies), and place names (strings). These similarity metrics are sometimes combined into workflows to conflate place entries.

Geographic Information Retrieval

Geographic information retrieval (GIR) refers to retrieving relevant geographic information according to users' demands or queries (Jones and Purves 2008). The data sources that are used for retrieving information are either structured or unstructured; nevertheless, existing research on GIR has usually focused on retrieving geographic information from unstructured data, especially on getting information from natural language texts on the Web (Purves et al. 2007).

It should be easy to get the inherent connection between GIR and geospatial semantics, for it has been estimated that 13–15% of Web queries contain place names (Jones et al. 2008). In order to retrieve relevant results, it is critical to understand the meaning of both the user queries and the candidate results (Janowicz et al. 2012).

Here, one important research issue in GIR is place name disambiguation (i.e. toponym disambiguation, in other words) which aims at understanding the actual geographic place that this place name really refers to: because it is a common sense that different places may have the same name (e.g. there many cities named "London" on the earth) and one place may have several different names (e.g., Beijing can also be called "Peking" or "Jingcheng"). Thus, how to identify the correct geographic place when a place name shows up in a query or in a Web document worth investigating, which has been studied based on a general strategy that measure the similarity between the current contexts of the place name (i.e., the surrounding words) and the likely contexts of each possible candidate place (Buscaldi and Rosso 2008).

Another important research issue in GIR is ranking candidates based on the input query. Such a topic often boils down to computing the matching score between the input query and a candidate result. Once the matching scores are computed, the candidates can then be ranked (Frontiera et al. 2008).

Geospatial Semantic Web and Linked Data

The Semantic Web is a vision in which the Web was populated with structured and semantically-annotated data that can be consumed by not only human users but also machines. It can be considered as an enhancement of the existing document-based Web on which the unstructured and natural language contents in Web pages are difficult for machines to understand.

Because the Semantic Web requires the current Web to be populated with structured and semantically-annotated data, Linked Data has been proposed by the World Wide Web Consortium (W3C) as a general guidance for publishing such data (Heath and Bizer 2011). The term Linked Data has a twofold meaning that is often used interchangeably. On the one hand, it refers to four principles for publishing well-structured data, such as using Uniform Resource Identifiers (URIs) and providing data descriptions readable to both humans and machines. On the other hand, it refers to the data that have been published following these four principles.

From 2007 to 2014, more than 570 datasets (with billions of RDF triples) had been published on the current Web (Hu and Janowicz 2016) which had formed the Linked Open Data (LOD) cloud that had been considered as a prototypical realization of the Semantic Web vision.

Place Semantics

It is obvious that place plays an indispensable role in human communication, and place names are frequently used in daily conversations (Winter and Freksa 2012; Cresswell 2014). Research on place semantics focuses on understanding the meaning of places through human descriptions and human-place interactions. Two types of place-related data can be found on the Web and have been studied. The first type contains only textual descriptions about places; the second type of data contains associations between descriptions and geographic coordinates.

Place semantics can be studied from thematic, spatial, and temporal perspectives. The thematic perspective examines human experiences toward places through natural language descriptions. The spatial perspective often focuses on representing the vague boundary of places or on spatially representing places based on the surrounding landmarks. The temporal perspective examines the time when people are more (or less) likely to interact with a place (Ye et al. 2011).

Cognitive Geographic Concepts and Qualitative Reasoning

Cognitive geographic concepts generally refer to the informal geographic knowledge that people acquire and accumulate during the interactions with the surrounding environment. Since not every GIS user has received formal training, understanding the conceptualization of general people towards geographic concepts can facilitate the design of geographic information systems.

The informal conceptualization of people on geographic concepts and spatial relations can be formally and computationally modeled to support qualitative reasoning. The term qualitative reasoning should be differentiated from the term qualitativeness which may imply descriptive rather than analytical methods (Egenhofer and Mark 1995).

6.2.5.2 Geospatial Semantic Relations on Land-Use/Cover Maps

Land Cover (LC) can be considered a geographically explicit feature that can be used in different disciplines (geography, ecology, geology, forestry, land policy and planning etc.) as a geographical reference (e.g. for land-use, climatic or ecological studies). In modern geographic databases, due to its relatively easy spatialization, LC has become a sort of "boundary object" between different disciplines. This, on the one hand, enhances the value of the LC information, but on the other hand, enlarges the base of potential users, posing new challenges for its harmonization and consistent use.

Many geographic entities have undertaken surveys of land cover (Jansen et al. 2008); often these have been made on the basis of a particular categorization system (often termed classification system) or class set (usually termed classification, nomenclature or legend). However, class definitions are often imprecise, ambiguous or, sometimes, absent. The build-up of the definition in the form of a narrative text is unsystematic and in any case do not always reflect the full extent, amount and type of the information, and is not logically structured. In addition, this structure of a database makes it very difficult to manage with modern GIS techniques. Semantic interoperability is actually the main challenge in Spatial Data Infrastructures. The representation of the multiplicity of the aspects summarizing a specific feature of the real world is drastically reduced or partially lost. Often some vagueness in the class definition is artificially included by the map producer to hide some "technical anomalies" when reproducing a certain feature on the map. Moreover, vagueness or extreme complexity in the class definition makes it difficult to correctly assess the accuracy of the data set. Semantic interoperability is the most evident and tangible result of this situation. Interoperability is defined as "the ability of systems to operate in conjunction on the exchange or re-use of available resources according to the intended use of their providers". In the case of "semantic interoperability", it refers to the understanding of the "meanings" of different classes and relations among concepts.

The graded and fuzzy nature of common LC categories derived from traditional classifications/legends has been recognized for a long time by the remote sensing community, however limited efforts to sort out semantic uncertainty in land cover studies has been proposed. Several studies to ascertain classes semantic similarity have been done and several techniques have been elaborated. However, especially when using a feature matching process to solve semantic discrepancies, it would be more effective to concentrate the efforts promoting and testing new ways to formalize the semantic of a classification system rather than continue to implicitly support obsolete paradigms (class names and narrative text description) that make the efforts for semantic interoperability difficult, time consuming and, in many cases, ineffective.

A LC classification system therefore should be able to support the many different ways a database can be generated and possess the following functions/advantages:

- be potentially able to converse with other systems. This inherent harmonization property should not rely (only) on expert judgment but the harmonization process should be automated as much as possible;
- recognize the balancing act inherent in classifying. A classification will never be able to fully represent all the aspects of the real world, and therefore it must be clear it reflects (just) a specific scope for which has been developed;
- render voice retrieval12 by allowing users to detail and compare classes using a detailed class description (systematically organized with a list of explicit measurable diagnostic attributes), thus avoiding the risk of systems being impermeable to the end users;
- standardization process should focus on the rules and conditions how a feature is conceptualized rather than acting just on the class name;
- the "formalization of the meaning" of the system and its components should be formulated with the most modern methods of modelling; and
- a modern classification should not be considered an isolated structure but more a functional component of a rather complete system for data management.

Figure 6.18 shows the framework of the LC classification system that have been used in China which basically abides by the above rules. The hierarchy in Fig. 6.18 is formed by the attributes of land use/cover. In other words, it presents the semantic relations among object types (simple or complex). These semantic relations may be grouped into three different categories: Is-a, Part-of and Member-of.

- Is-a relation: an IS-A relation exist between a super-type and a sub-type, which means a sub-type can derive from its super-type. A sub-type inherits the properties of the super-type, and can also define its own additional properties, whereas a super-type cannot. For example, in Fig. 6.18, cultivated land is a super-type of irrigable land. Irrigable land can inherit the properties from cultivated land, and it

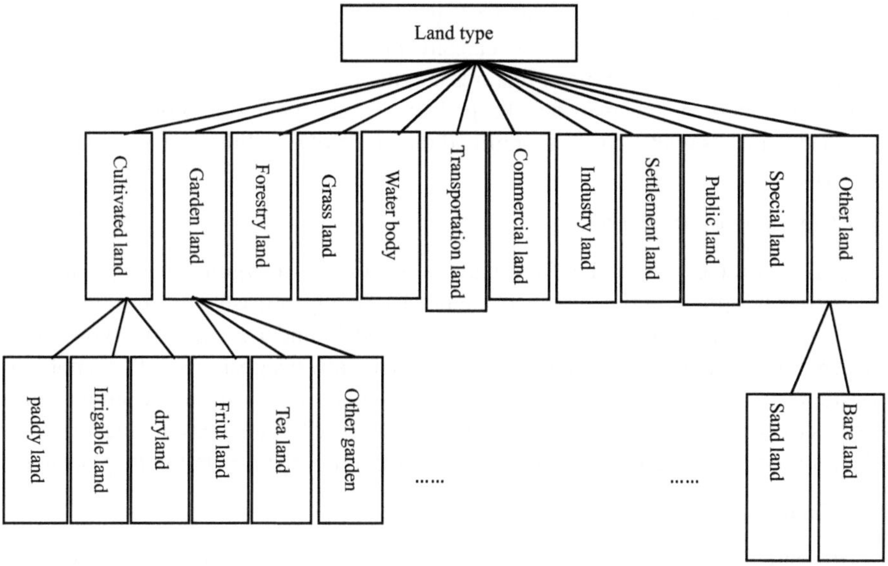

Fig. 6.18 Land cover/use classification system used in China: a semantic hierarchical tree

can also define a number of additional properties such as amount of water used per season but cultivated land cannot define additional properties.

- Part-of relation: a Part-of relation exists between a higher order object and a lower order object which means a higher object can be formed by a lower object that belong to a different classification hierarchy. A higher order object may inherit attributes from a lower object.
- Member-of relation: a Member-of relation exists between an object and an association of a number of the same type of objects which means that a given object is a part of the association of the same object.

6.3 Principles/Rules in Land Use/Cover Map Generalization

The following principles should be obeyed in land use/cover map generalization:

(1) Spatial accuracy

Pay attention to spatial locations of areal land patches. Borders of areal patches should be determined under the control of the plan and elevation control networks so that the spatial accuracy of areal patches can meet the requirements. The features (such as waterbodies, roads, buildings, boundaries of political regions, and other features closely related to land use/cover) collected from topographic maps should be selected as accurate as possible; while the other features (such as annotations and elevation points) can be deleted or retained some of them if really necessary.

The more important a type of features are, the more probable this type of features can be retained on the map and keep their original positions. For example, if a railroad conflicts with a river, the railroad may be moved while the river may not be moved.

(2) Closeness of areal patches

It is a common sense that land patches are closed polygons on maps, and the edges of each polygon (i.e. an areal land patch) usually consists of at least more than two types of land boundaries which means the boundaries of a polygon are in different layers in the map database. Thus, it is not always guaranteed that a polygon for representing a corresponding land patch is really a closed one. Therefore, it is necessary to check the topological relations among the boundaries and guarantee that their topological neighbouring relations are correctly recorded in the map database.

(3) Retainment and deletion

Pay attention to the shape and size of land patches to avoid arbitrarily deleting or adding land patches, to avoid changing a curve into a straight line, to avoid changing a thin polygon into a wide one etc. Retain the linear objects (e.g. rivers, a line of forest, roads, irrigable canals and ditches) whose width is less than given threshold values and delete the ones whose width is greater than given threshold values. As far as small areal patches are concerned, a threshold value should be given for the purpose of keeping the landscape not too broken.

(4) Amalgamation of areal land patches

The areal land patches that are topologically neighbouring each other and semantically similar can be merge to form a new patch. For example, a patch of paddy field and a patch of pond which are neighbouring on the map with each other can be merge.

(5) Encoding of features

All objects on the map should be encoded and the code assign to each object should be excusive.

(6) Generalization of features

Generalization here means simplification of map features in shape, size and the number of objects. It is different from selection. Generalization results should be as similar as possible to its original features in distribution characters.

(7) Consideration of regional characters

Threshold values and indices used in map features selection should consider geographical regional differences or characters. For example, a pond of 1 hectare is usually deleted from the map at scale 1:50 K if it is in a region of lakes and ponds; whereas it should be retained if it is in an arid region.

(8) Super-type priority

If an object is too small in size to be retained on the map, it can be mere into its super-type.

(9) Consideration of uniqueness

If an object is unique and very important, it should be retained on the map though it is too small in size to be directly symbolized on the map.

6.4 Generalization of Continuous Areal Features

There are majorly three categories of approaches for land use/cover map generalization: (1) rule-based approaches, (2) approaches that consider maintenance of spatial relations, and (3) agent-based approaches.

6.4.1 Rule-Based Land Cover Map Generalization Approaches

Land cover map belongs to categorical maps that usually represent different types of land covers using different colors or textures. When the scale of a land cover map becomes smaller, the land patched and linear features on the map need to be generalized so that they can be displayed clearly. Many scholars have researched on this issue and made a lot of achievements (Gao et al. 2004; Gao and Gong 2005). In fact, generalization of land cover/use maps is a complicated process which not only need to abide by fundamental principles and rules in map generalization, but also need to consider the distribution characteristics of types of land covers and the purpose of the map under different conditions (e.g. target map scale and map users). In addition, land covers are closely related to natural and social factors, such as soil type, climate and land use type; thus, it is necessary to consider these factors when the boundaries of land covers are determined. In this sense, these factors should not only be integrated in the process of land cover map generalization, but also should be taken as rules in evaluating generalization results.

Map generalization is always based on rules. Here, map generalization rules can be explained in the either narrow or broad sense. In the broad sense, map generalization rule refers to all rules and principles that should abide by in the process of map data generalization, and some scholars gave it another name called "map generalization knowledge" (Buttenfield and McMaster 1991). In the narrow sense, map generalization rules specially means the rules generated by the "production rule method" borrowed from artificial intelligence which usually uses "IF-THEN" structures for representing knowledge.

6.4.1.1 Representation and Application of Map Generalization Rules

There are many methods for representing map generalization rules, such as production, logic, semantic network, frame, state space, conceptual dependency, script, Petri etc. Here, only the production rule method is employed, because: (1) the rules expressed by this method are natural, concise and easy to understand; and (2) the rules generated by the production rule method are independent, therefore deletion/addition of a rule does not affect other rules, which makes the changes of the rule library easy and simple (Shea and McMaster 1989).

A typical syntactic structure of the production rule method is as follows:

```
IF <conditions> THEN (behaviors/results)
```

This structure consists of two parts: the conditions and the behaviors/results. The left part is a conditional sentence which describes the states of the generalized objects; the right part consists of one operation/result or a group of generalization operations/results. In other words, the left shows the states or conditions of the task and the right shows the methods for carrying out the task or the results of the task.

To put production rules into practical use, a production rule library and a reasoning rule library should be established. The production rule library consists of production rules, and the reasoning rule library consists of the factors and methods for generalizing map features. When the map generalization system is executed, if the fact (i.e. a will-be-generalized object or a group of will-be-generalized objects) matches a condition of the production rule library, this rule is applicable and should be marked as "applicable". However, this rule is not executed at once. Go on searching the other facts and mark the ones that are applicable. After all of the facts have been searched and all applicable facts are marked, compare all of the applicable facts and find the most appropriate one and execute it behaviors, and then continue this process by executing the second most appropriate one.

6.4.1.2 Rule for Land Cover Map Generalization

Because land covers on maps are represented using areal symbols, this section mainly focuses on areal land patches. The following paragraphs present the major rules for land cover map generalization, and some rules are represented using pseudo codes.

Rules for Representing the Relations Between Land Use and Land Types

Determine the thematic types of the land covers according to the purpose of the target map and then determine the importance degree of each type of land cover. There are many methods for classifying vegetations (i.e. the land cover). A typical one used in vegetation classification is: vegetation type group, vegetation type, vegetation sub-type, formation group, formation, sub-formation, association group, association, sub-association. As far as a vegetation map is concerned, the type of land covers that can be represented on the map relies on the types of vegetation save in the vegetation database; therefore the vegetation classification system should be considered to know if it can meet the basic mapping requirements before the map is made.

Continuity Rule of Land Covers

Vegetations are generally continuously distributed in the geographic space, and there exist tall trees, low trees, tall shrubs, low shrubs, grassland, lichen land, bare land. Generally, no grassland appears between a patch of land covered by trees and another patch of land cover by shrubs except for human's interruption.

If a long, narrow patch of shrubs appears between a patch of land cover by trees and another patches grassland and the width of the shrub land is less than a given threshold, the shrub land usually may not be deleted. On the contrary, its area should be exaggeratedly displayed on the generalized map for the purpose of showing the continuity of the vegetations. This rule also manifest that the continuity rule is usually used together with some geometric rules.

Rule for Generalizing the Attributes of Land Covers

Land cover maps belong to thematic maps. Generalization of thematic maps needs to consider thematic information; therefore, generalization of attributes is of great important in land cover map generalization. Generalization of attributes in land cover map generalization refers to transformation of the types of land cover, i.e. how to change detailed types of land covers to coarser ones. It is apparent that this change is irreversible. In other words, types of land covers can be transformed from detailed ones to coarser ones but not vice versa. Meanwhile, it should be noticed that the thematic land covers must be displayed in more detail than other ones.

Figure 6.19 shows a vegetational classification system ever used in Hongkong which is basically the same as a land cover classification system. This classification system, which has tree grades of classes, has been employed in the establishment of Hongkong's vegetation database at scale 1:20 K. The land patches saved in the database belong to the third grade of classes.

Logical consistency in vegetational type transformation means when the type of land patches at the third grade is transformed into that at the second grade, the type of a patch of artificial trees may not be transformed into a patch of grassland or shrubs, it can only belong to trees. In other words, a type at the second grade should be the father of some types at the third grade. The transformation must guarantee the inheritance relations between the son and the father.

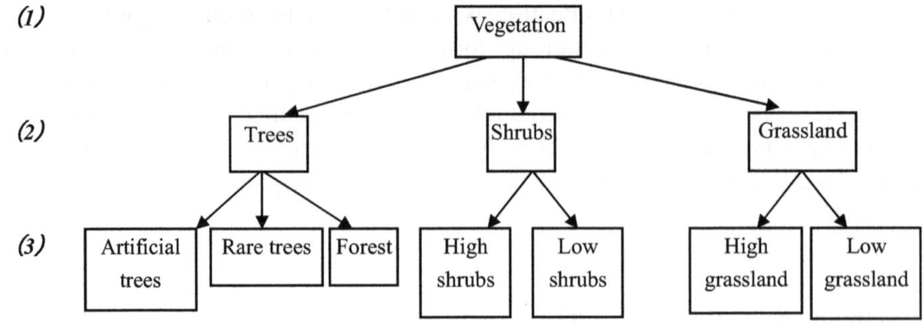

Fig. 6.19 A vegetational classification system used in Hongkong

Of course, if distribution of trees is the theme of the map, the third grade of the types of trees can be retained on the target map so that the theme can be displayed in detail, while the other types of vegetations should still be generalized to the second grade.

The idea of this transformation can be expressed by the following rule:

```
   IF NewClassGrade (Objecti) is lower than OldClassGrade (Objectj) AND
NewClassType (Objecti) is parent class of OldClassType (Objectj)
   THEN Transformation Type of objecti to newClassType)
Where,
   NewClassGrade (Objecti) —grade of the new type of the vegetation
   OldClassGrade (Objectj) —grade of the old type of the vegetation
   NewClassType (Objecti) — new type of the vegetation
   OldClassType (Objectj) —old type of the vegetation
```

Rules for Making Maps

Making map is related to a wide range of contents, from symbolization of singular map features to beautifying map surface. Here, only the conditions that do not meet the basic requirements of making maps in representing attribute information after map generalization are discussed.

When the type of a land patch is transformed from a sub-type (i.e. child-type) to its super-type (i.e. father-type), some topologically adjacent land patches may become belonging to the same type, which is not allowed in making map; therefore, the two land patches should be merge into one patch. So, after land type transformation, it is a necessary step to check if any two topologically adjacent land patches belong to the same type of land cover.

This rule can be expressed as follows:

```
IF ClassType (Object_i) = ClassType (Object_j) AND Object_i is adjacent to
Object_j
  THEN Merge the two Objects (Object_i, Object_j)
```

Geometric Rules

Resolving power of human eyes has a minimum value (it is usually a length value, say MIN_l). Only if the size of a geometric symbol on the map is greater than MIN_l, can this symbol be discerned by human eyes. One of the major objectives of map generalization is to avoid overlap or/and conflict of map symbols, and in the meanwhile keep the map legible and the topological relations among map features correct. Usually, the rules at the following three aspects need to be considered.

(1) Area threshold

Experiences from pioneering cartographers tell us that if the area of a polygon on the map is less than 0.5 mm^2, the polygon can not be discerned by human eyes (Muller and Wang 1992); hence, any land patch whose area is less than 0.5 mm^2 should be taken into consideration in the process of map generalization. More importantly, the land patches should be processed according to the theme of the target map so that the theme can be salient and clear. For example, if the theme of the map is forest, a small land patch whose size is less than 0.5 mm^2 may be processed using the following strategy:

a: if this land patch is forest and there is no other forest land around it, the small land patch should be retained on the target map and its size should be enlarged.

This rule can be expressed as follows:

```
IF Area (Object_i) < Area threshold of Forest AND.
There is no forest parcels around Object_i.
THEN Enlargement Operation (Object_i)
```

b: if the enlarged land patch has a conflict with its neighbouring roads, rivers or boundaries, the land patch may be moved before it can be enlarged.
c: if this forest land has other neighbouring patches of forest land, it can be merged or amalgamated into its neihbouring patches of forest land to form a new big land patch.
d: if this small land patch is not forest, it can be merged or amalgamated into its neighbouring land patches on condition that this operation does not affect the continuity of the land cover.

(2) Distance threshold

This threshold can be the shortest discernable distance between two topologically separated objects on maps, and it also can be the smallest width of a shape on a polygonal object.

If the smallest distance between two objects is less than the distance threshold, one of the objects who is less important on the target map can be moved. This rule may be described as follows:

IF Distance ($Object_i$, $Object_j$) < Threshold and ($Object_i$) is more important than ($Object_j$)
THEN move ($Object_j$)

If the width of a shape on a polygonal object is smaller than the distance threshold, this small shape should be deleted and its area should be divided by its original neighbours. This rule may be described as follows:

IF Width ($Shape_i$) < Threshold
THEN Delete ($Shape_i$) and Divide ($Shape_i$)

(3) Topological relations

Generalization of map features can cause the change of topological relations among map objects. Changes of topological relations should abide by many rules; otherwise, the changes are not acceptable. For example, in Fig. 6.20, some topologically connected areal vegetation land are generalized. The area of the smallest land patch should be delete. However, after the deletion operation, the smallest land patch should be divided by its neighouring land patches (Fig. 6.20b) but not leave a hole on the map (Fig. 6.20c). Another example is the merge of two areal land patches, shown in Fig. 6.21. The two patches are topologically adjacent (Fig. 6.21a); after they are generalized, they should become a polygon (Fig. 6.21b). Otherwise, if they become topologically contained (Fig. 6.21c, e or separated (Fig. 6.21d), the resulting map is not acceptable.

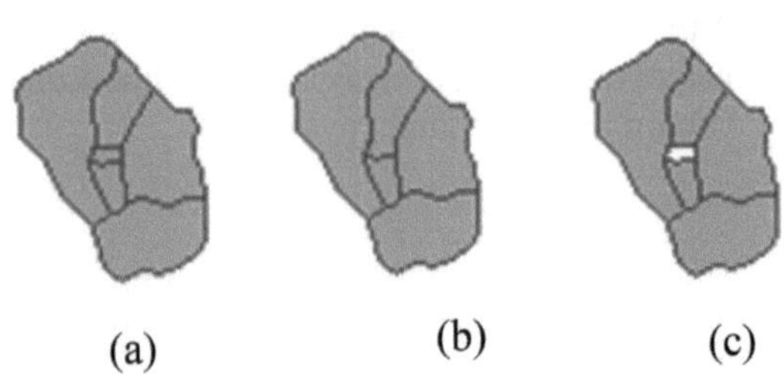

Fig. 6.20 Amalgamation of a small polygon: (**a**) source map, (**b**) correct map after the deletion operation, and (**c**) wrong map after the deletion operation

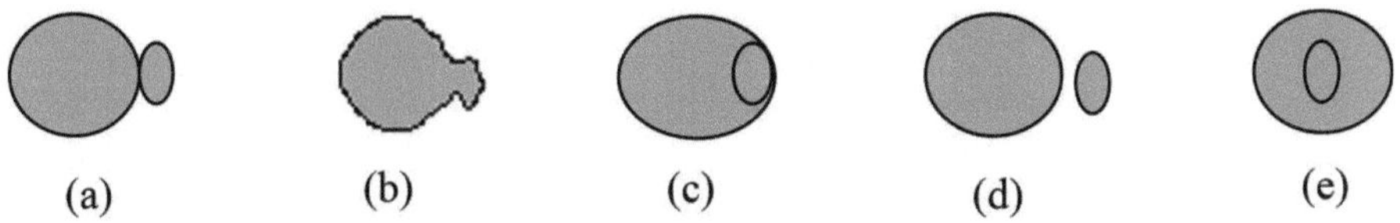

Fig. 6.21 Merge of two adjacent polygons: (**a**) source map, (**b**) correct map after the merge operation, and (**c–e**) are three incorrect results after the merge operation

Rules for Balancing the Area Ration of the Types of Lands

Although it is important to stress those types of land covers who are the theme on the target map, it is also noticeable to keep balance among the area ratios of types of land covers before and after map generalization, which can truly express the distributional characteristics of the land covers at this region and thus generate a reasonable target map.

Rules for Coordinating the Relations Among Land Covers and Other Geographic Features

Generally, distribution of land covers is closely related to other geographic factors such soil type. For example, when the terrain of a region changes sharply, the land surface usually changes from soil to rocks, and the types of vegetations also naturally change. Thus, the change in vegetation caused by the change of soil should be considered in map generalization. In other words, the contexts (i.e. other factors) should be taken into account when vegetation data is generalized.

Figure 6.22 shows an example of a vegetation map (i.e. land use/cover map) generalization based on rules.

6.4.2 Areal Patches Generalization Approaches Considering Spatial Relations

As far as the combination of two polygonal objects on the map is concerned, if the semantics of the two objects are the same, the combination is called aggregation; otherwise, if the semantics are different the combination is called amalgamation (Shea and McMaster 1989). So, an aggregation operation combines two polygons with the same semantics but the two polygons are not definitely topological adjacent; thus, when they are combined there usually heterogeneous lands appearing between them; whereas an amalgamation operation combines two topologically adjacent but semantically different polygons.

On the one hand, it is obvious that a land patch must have a topological neighbour except that it is an isolated island; however, it is uncertain that a land patch has a

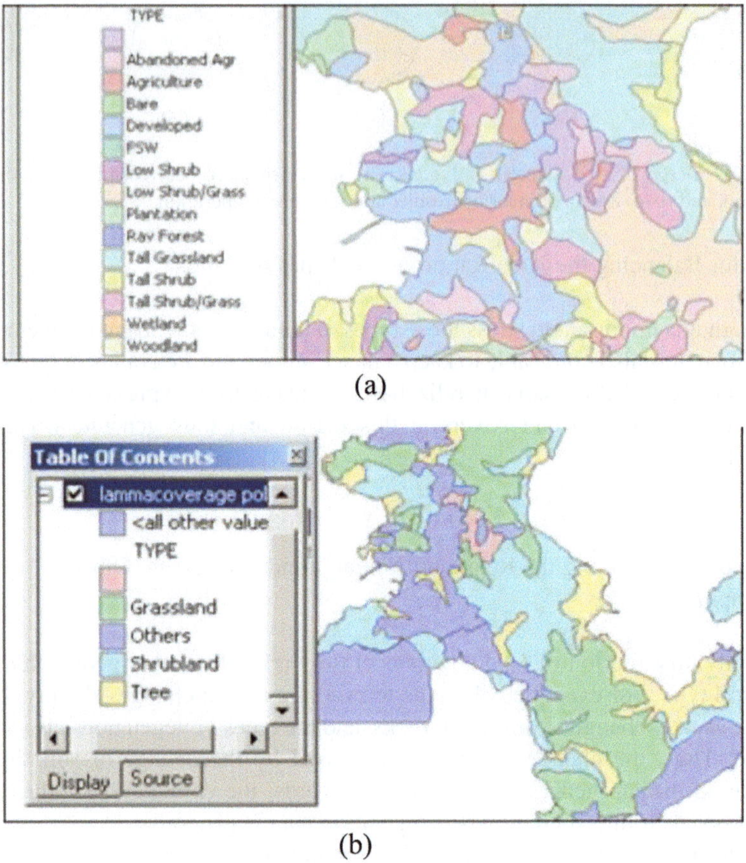

Fig. 6.22 An example of land cover map generalization based on rules. The two maps are not shown to the scale exactly. (**a**) Source map at scale 1:20K. (**b**) generalized map at scale 1:50K.

neighbour that is semantically similar with it. On the other hand, generalization of land use/cover map mainly deals with small patches of land. Therefore, generalization of a small land patch (say, a) may be discussed according to the following three conditions:

- If a does not have a neighbor who is semantical similar with it and the area of a is smaller than its topological neighbours, a should be amalgamated into its topological neighbours, i.e. a is divided by its neighbours and the type of a is changed after map generalization.
- If a does not have a neighbor who is semantical similar with it and the area of a is greater than at least one of its topological neighbours, it is possible that a amalgamates with its neighbours. To be exactly, a may occupy the area of its neighbours and it keeps its original land cover type.

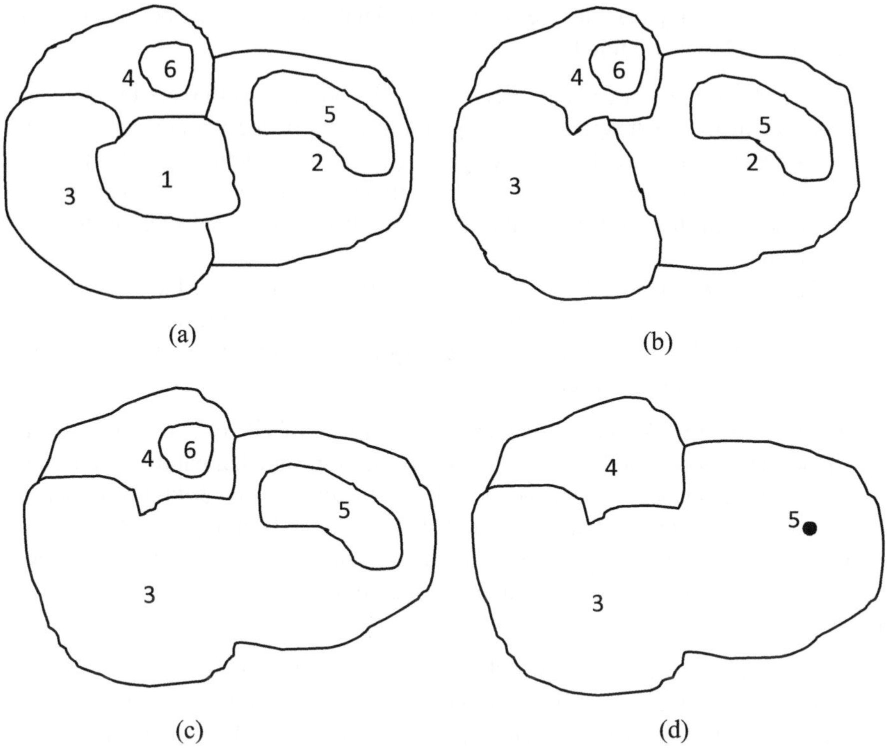

Fig. 6.23 Change of topological relations in land patches generalization. (**a**) Source map. (**b**) After amalgamation. (**c**) After aggregation. (**d**) After deletion and collapse

- If a has a neighbor who is semantical similar with it, it is possible that a may aggregate with its semantical similar neighbours and divide the small areal patches between them.

Besides aggregation and amalgamation, collapse and deletion are also commonly used operations in areal land patch generalization (Ai and Liu 2002; Liu et al. 2003), shown in Fig. 6.23. The following discussion does not pay attention to the algorithms of the four operations but focuses on the maintenance/check of spatial relations in the process of land patches generalization, for it is more complicated and uneasy to implement and therefore is the key in those algorithms, supposing that the topological relations among land patches have been got, including the topological adjacency relations among land patches, topological joint relations among land patches and the edges/arcs, and topological neighbouring relations among arcs/edges.

(1) if a small areal land patch (say, P) has disappeared after an amalgamation operation, the topological relation among all of the areal land patches that are

topologically adjacent to the small areal land patch should be checked and maintained, which includes

① Search for the source data to get the areal land patches that are topologically adjacent to P. Change the corresponding items of the variable of P defined using the structure PatchDescription (it has been defined in the previous section of this chapter). The items include the number of the land patches adjacent to this land patch, the IDs of the land patches adjacent to this land patch etc.
② Take the areal land patches that are topologically adjacent to P one by one, and change the coordinate chain of each patch. The change of coordinates is caused by the amalgamation of the polygons (i.e. the areal land patches that are topologically adjacent to P).

For example, Fig. 6.23b is the amalgamation result of Fig. 6.23a, in which patch 1 is divided by patch 2 and patch 3. Obviously, after the amalgamation operation the adjacency relations between patches 1 and 2 and patches 1 and 3 disappear; and the boundary coordinates of patches 2 and 3 are changed; and patch 2 and patch 3 become neighbours (they are not neighbours in the source map). Here, patch 4 does not play a role in the amalgamation operation, for its semantics is not similar to patches 1, 2 and 3.

(2) if a small areal land patch (say, P) has disappeared after an aggregation operation, the topological relation among all of the areal land patches that are topologically adjacent to the small areal land patch should also be checked and maintained.

① the aggregation usually results in a bigger land patch which inherits the containment relations and adjacency relations that P has with other land patches; thus, the corresponding items of the land patches whose topological relations have changed due to the aggregation should be changed in the variables defined using the structure PatchDescription.
② the coordinates of the boundary of the bigger land patch should be changed according to the aggregation results.

Figure 6.23c is the aggregation result of Fig. 6.23b. After the aggregation operation, patch 2 has disappeared, and patch 5 contained by patch 2 become contained by patch 3, and the patches that are adjacent with patch 2 in Fig. 6.23b become adjacent with patch 3.

(3) if a small land patch is deleted, the containment relation that is recorded in the land patch that contains this small land patch should be deleted, and the area of the small land patch should be possessed by the land patch that contains the small land patch.
(4) if a land patch has collapsed, its topological relations with other land patches usually do not change.

For example, patch 5 in Fig. 6.23c has become a point symbol in Fig. 6.23d. however, its containment relation with patch 3 does not change.

6.4.3 Agent-Based Approaches for Areal Patches Generalization

6.4.3.1 Basic Concepts of Agent

In computer science, a software agent is a computer program that acts for a user or other program in a relationship of agency: an agreement to act on one's behalf. Such "action on behalf of" implies the authority to decide which, if any, action is appropriate (Nwana 1996; Schermer 2007).

An autonomous software agent has the following attributes:

- it is not strictly invoked for a task, but activate itself;
- it may reside in wait status on a host, perceiving context;
- it may get to run status on a host upon starting conditions;
- it does not require interaction of user; and
- it may invoke other tasks including communication.

The concept of an agent provides a convenient and powerful way to describe a complex software entity that is capable of acting with a certain degree of autonomy in order to accomplish tasks on behalf of its host. But unlike objects, which are defined in terms of methods and attributes, an agent is defined in terms of its behavior.

All agents are programs, but not all programs are agents. Four key notions that may distinguish agents from arbitrary programs: reaction to the environment, autonomy, goal-orientation and persistence (Wooldridge and Jennings 1995; Wooldridge 2002). There are four essential types of intelligent software agents: (1) buyer agents or shopping bots, (2) user or personal agents, (3) monitoring-and-surveillance agents, and (4) data-mining agents.

6.4.3.2 Role of Agents in Polygonal Map Generalization

Agent technique has been used in map generalization for years. Qian and Wu (2004) proposed an agent-based monitoring model for automated map generalization (Fig. 6.24) which integrates agents into the procedures of map generalization and can monitor the whole process of map feature generalization. This procedure is as follows: the agents produce the electronic maps using the source geo-spatial databases by map generalization knowledge and rules. In this process, the agent working in the host computer must know the basic information of the agents generated by itself, including the number, position, user name and time, current status of the generated agents. In the meanwhile, the generated agents also need to know the information of the agent working in the host computer, including the position, user name and current status (Fig. 6.25).

Galanda (2003) proposed an agent-based hierarchical model for polygon generalization. The agents constructed in this model are classified into four categories and

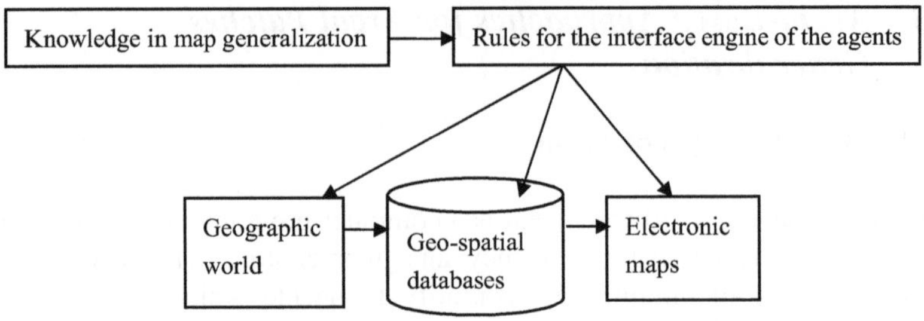

Fig. 6.24 Principle for agent-based map generalization

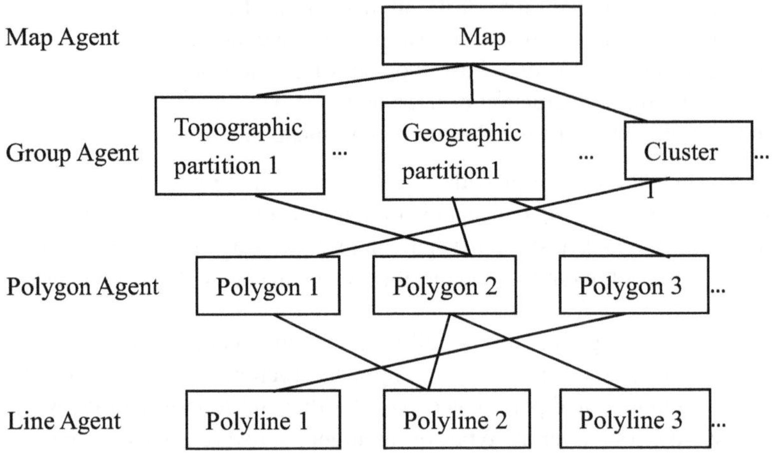

Fig. 6.25 An agent-based hierarchical structure used in land patches generalization

are set at different levels in light of their functions. They are called map agent, group agent, polygon agent and line agent, respectively (Fig. 6.26).

Map agent: Every map has one map agent. It is responsible for constraints and generalization operations concerning the whole polygon map and the instantiation of the group agents at run time.

Group agent: Group agents handle contextual generalization, i.e. conflicts between polygon objects. They are composed of several objects sharing a common geometric, topological or semantic relation. Thus, reasonable group agents may be attached to a cluster, an alignment, a category, a topological partition or a geographic partition (e.g. a collection of polygons bounded by rivers or roads). A non-exhaustive list of group agents may include cluster agents, category agents, geographic partition agents and pattern agents.

Group agents can recursively subdivide themselves if needed, that is, a group agent can spawn off other group agents besides the polygon agents. Furthermore, group agents may be built upon arbitrary combinations of the listed types of group agents.

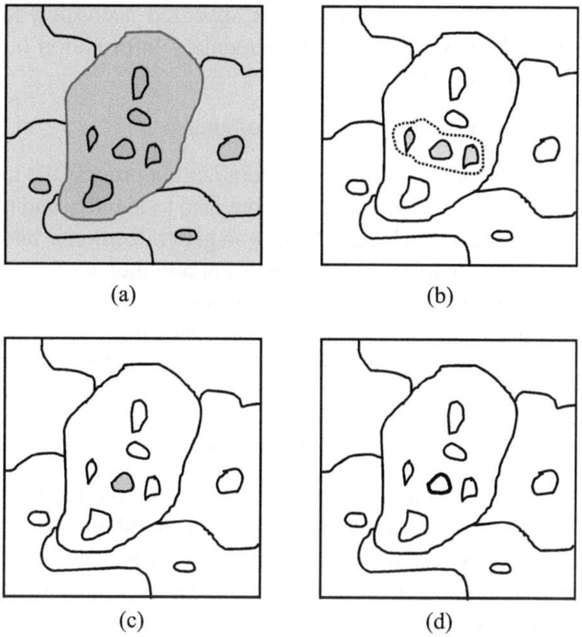

Fig. 6.26 Agents at different levels (**a**) Map agent (**b**) Group agent (**c**) Polygon agent (**d**) Polyline agent

Polygon agent: A polygon agent coordinates the generalization of an area object. Constraints and operations acting on an individual polygon are evaluated and performed without considering the agent's spatial context (e.g. enlargement). However, the geometric transformation of one polygon induces always at least the modification of one other polygon. So, every change must be propagated to those objects sharing a common geometric primitive with the transformed object. In other words, side-effects of independent generalization have to be settled.

Line agent: Line agents are delegated to polylines bordering a polygon object and their generalization (e.g. simplification). Exactly two polygon agents can supervise a line agent of the first generation. Similar to group agents, they are enabled to recursively subdivide themselves in order to perform generalization on homogenous line parts.

6.4.3.3 Agent Life Cycle in Polygon Map Generalization

The generalization process for polygonal data consists of three main stages. A pre-processing stage prepares the database for the second stage, the iterative agent-based generalization. Finally, the user undertakes an evaluation of automated generalization's result during the third stage.

- Pre-processing: data specification and analysis

First of all, constraints are specified according to the basic conditions of the generalization task and supplementary information (auxiliary data) is calculated on the database objects.

- Generalization: life cycles of agents

At the beginning of the generalization process the map agent is activated. It has to care about its own constraints but also to identify and trigger group agents on the fly. Procedural knowledge of polygon generalization is needed to decide either to try first to satisfy a parent agent's constraint and then work on the child agent's constraint or to first initialize the child agent's generalization and then the parent's constraints. In practice, a mixture of these approaches may be applied. Again, procedural knowledge can help choosing the child agent to be triggered first amongst all the child agents.

- Post processing: final evaluation

After the map agent's life cycle is completed, the user evaluates the final result with the help of a detailed report (severity of remaining conflicts, happiness per agent or category etc.) output automatically by the system (Ruas 2001). If necessary some interactive re-generalization is performed.

The process of improving an agent's happiness starts when its parent agent changes its state to active. The following sequence of constraints evaluation, proposing plans, triggering the best plan and re-evaluation is the same generic behavior of all agents and called an agent's life cycle. The life cycle ends when either all constraints are satisfied completely (perfect state) or no plan to be tried is left. In any case the database is updated by the best state ever reached in the life cycle. In other words, the plan that leads to the perfect or best state is triggered in any case sooner or later, but 'best plan' acts as a heuristic to find the best state more efficiently. The agent is set to passive again, that is, the control of the generalization process is returned to the parent agent.

6.4.4 A Vector and Raster-Based Approach for Categorical Map Generalization

Categorical maps, including maps (or databases) of soil, geology, vegetation, or classified remote sensing images, are a type of maps that are frequently used in GIS and thematic cartography. Networks of political or administrative boundaries can also be considered a special case of this maps type. Categorical maps are commonly modeled as either vector data (i.e., as polygonal maps or polygonal subdivisions) or as raster data. Raster categorical data mainly originate from grid samples, remote sensing imagery, or interpolated and classified point samples. Vector data are usually digitized from the corresponding categorical maps. Although there are tools available in current commercial GIS and cartography systems that allow processing raster and vector categorical data for purposes of analysis and display, specific methods for automated generalization of such data are still necessary to develop. In most of the literature, for vector categorical maps line generalization algorithms are used instead of polygon-oriented methods, and for raster maps generalization of raster data sets are essentially equivalent to simple pixel-based image processing operations, not respecting the object nature of "raster polygons" (regions, connected components). Here, a more sophisticated method considering the preliminary nature of both raster and vector data in categorical maps is discussed (Peter and Weibel 1999).

The method aims at efficiently integrating vector and raster-based techniques into a comprehensive generalization methodology. To achieve this objective, three elements are looked at:

(1) The first element concerns the specification of so-called generalization constraints, that is, conditions of geometric, topological, semantic, and Gestalt nature, which govern the process of categorical map generalization.
(2) The second element of an integrated methodology has to do with the translation of the constraints into tools for assessing the need and the effect or quality of generalization (assessment tools, measures) and tools for achieving the necessary generalization transformations (transformation tools, generalization algorithms).
(3) The third component relates to the conversion between raster and vector data in both directions.

6.4.4.1 Constraints to Categorical Map Data

A constraint in the context of generalization can be defined as a design specification to which the solutions to a generalization problem should adhere (Weibel and Dutton 1998). A constraint is meant to limit the number of possible solutions without binding it to a particular action. Constraints can be classified according to their functions in map generalization, which seems appropriate for generalization problems in a digital environment (Weibel 1996; Weibel and Dutton 1998), and four types of constraints are distinguished:

(1) Graphic constraints mainly deal with aspects of perceptibility such as size, width and separability. Categorical data consist entirely of area features where each category is assigned a different color fill. Depending on the color, different parameters for minimum perceptibility may apply.
(2) Topological constraints deal with basic topological relationships like connectivity, adjacency and containment, which should be maintained when generalizing data. Self-intersection and overlapping patches are also issues related to topology which occur as a result of erroneous generalization algorithms.
(3) Structural constraints define criteria that describe spatial and semantic properties of the data. Spatial structural constraints deal mainly with the preservation of typical shapes (on the patch level) or with the preservation of patterns and

alignments if multiple patches are involved. Semantic structural constraints deal with the preservation of the logical context of patches.

(4) Gestalt constraints relate to aesthetic aspects. These include the preservation of the patch characteristics as well as the retention of the overall visual balance when multiple patches or the whole dataset is considered. Gestalt constraints are complex and difficult to formalize for use in digital systems but nevertheless important since they represent aspects of cartographic knowledge

Constraints can also be classified according to their spatial application scope, which can either be a single patch, all patches of a category or a group of patches, a partition of the map or the whole map respectively. It should be pointed out that constraints for single patches are mainly related to the selection process rather than to the actual transformation, therefore they are not discussed here.

- **Constraints Related to Patches**

1. Minimum size (graphical): patches, which are too small, can be either deleted or enlarged.

2a. Minimum distance (graphical, vector): the distance between consecutive vertices of a polygon outline should not be less than the minimum visual separability distance.

2b. Minimum distance (graphical, raster): the distance between any parallel edges of the outlines (horizontal or vertical) of a region should not be less than the minimum visual separability distance.

2c. Self-coalescence (graphical, vector): the distance between any vertices of a polygon outline should not be less than the minimum visual separability distance.

3. Separability (graphical): the distance between two patches should not be less than the minimum visual separability distance.
4. Separation (topological): avoid separation of patches when deleting parts of it.
5. Islands (topological): patches, which can be identified as islands may be deleted or enlarged but should not be amalgamated with other patches of the same category.
6. Self-intersection (topological): avoid introduction of self-intersection of patch outlines.
7. Amalgamation (structural): disjoint patches of the same category may be amalgamated.
8. Collapsibility (structural): the area of eliminated patches should be distributed among the neighboring patches.
9. Shape/Angularity (structural): respect the global shape and angularity of patches.

- **Constraints Related to Categories**

10. Size ratio (structural): respect the size ratio for each category relative to the total area.

11. Shape/Angularity (structural): respect typical shapes and angularity of patches of each category.
12. Size distribution (structural): respect the given size distribution of patches for each category.
13. Alignment/Pattern (Gestalt): preserve typical alignments and patterns of patches of a category.

Constraints Related to Partitions or Groups of Patches

14. Neighborhood relations (topological): preserve given neighborhood relations.
15. Spatial context (structural): avoid introduction of illogical neighborhood relations (e.g., house in a lake).
16. Aggregability (structural): allow aggregation of categories if required and suitable super-categories exist.
17. Auxiliary data (structural): observe constraints imposed by auxiliary data (e.g., roads, rivers, point features).
18. Alignment/Pattern (Gestalt): preserve typical alignments and patterns of patches within the map or within a group of patches.
19. Visual balance (Gestalt): avoid gross changes in shape and distribution of patches, unless required by extreme scale change.
20. Equal treatment (Gestalt): ensure equal treatment within a partition of the map and avoid highly unequal treatment across all partitions.

6.4.4.2 From Constraints to Measures and Generalization Algorithms

Measures are the basis for formal descriptions of relevant characteristics of geographical entities at the patch, category and map level. A measure can be a simple formula (e.g., area calculation) or a complex algorithm, which may even require the computation of auxiliary data structures like a Delaunay triangulation. Measures can be either absolute (intrinsic), meaning that they can be interpreted and applied according to the analysis of one state of the database or be relative (extrinsic), meaning that measurements of two states of the database have to be compared and evaluated to decide if a solution is acceptable or has to be rejected. Most measures exist for vector and raster data but employ different methods for their computation. Many classes of measures can be distinguished according to the main characteristic they represent, e.g. size measures, distance and proximity measures, shape measures, topological measures, density and distribution measures, pattern and alignment measures etc.

Translating constraints to measures is a complex and crucial process within a constraint-based generalization system. The underlying concepts of constraints and measures are quite different. The goal of a constraint is to limit the number of acceptable solutions to a problem without binding it to a particular action. A measure on the other hand is a formal mathematical concept, which makes use of clearly defined formulae or algorithms. Only few constraints, e.g. minimal size, can be translated to a measure on a 1:1 basis. Most concepts, such as shape or visual balance are rather fuzzy and ill-defined terms. Hence, it is almost impossible to formally describe all properties that characterize such a spatial concept comprehensively. Translating constraints to measures is therefore also a selection process. The goal of the process is to make the main properties of a spatial entity available to formal mathematical descriptions. The degree to which this goal can be achieved has a major influence on the results of the generalization process. Generalization algorithms cannot deal with properties of patches or spatial entities that have not been formalized, nor can changes be evaluated for possible rejection of solutions. In general, graphical constraints can be formalized more easily and precisely than structural or Gestalt constraints.

6.4.4.3 Conversion Between Raster and Vector Data

Many achievements have been made on the methods by which vector data is converted to the raster structure, then generalized and finally transformed back to the vector model (e.g. Jaakkola 1998; de Berg et al. 1998) (in brief, vector-raster-vector). In general, this strategy makes more sense than the reverse (raster-vector-raster) since most raster operators, especially those involving neighborhood and contextual operations, are simpler and easier to implement than their counterparts in the vector domain. This may compensate for the relative loss of precision and semantic expressiveness that occurs when data is transformed. When data is reconverted after generalization, the source data (vector) might help with the interpretation of possible ambiguities that arise in the raster to vector phase.

No convincing argument, however, can be thought of that would justify a generalization strategy where raster data is converted to the vector model for generalization with subsequent re-transformation back to raster structure. When a generalization strategy involving bi-directional conversion is implemented, various effects and problems have to be analyzed. Unavoidably, transformation of data between continuous and discrete reference systems and vice versa results in a loss of information and/or precision. Piwowar et al. (1990) have implemented several conversion algorithms and have evaluated them based on qualitative, quantitative and efficiency criteria. None of them could satisfy all requirements at the same time. Since bi-directional conversion usually doubles the effects mentioned and these effects cannot be easily controlled, it should not be recommended for general use despite the fact that some generalization algorithms might be easier to implement in a raster environment. In addition, parameters for vector to raster conversion (e.g., sampling interval) are usually defined globally for the whole dataset. This may result in partial or complete loss of important local geometric information that cannot be taken care of during the generalization process. Thus, some scholars proposed to use *local* conversion to get a solution for this kind of problem (Peter and Weibel 1999), which will be discussed in this section later.

It is necessary to distinguish between the source data representation and the target data representation of a map. Commonly, one should tend to maintain the representation of the source data. In other words, conversion is to be avoided unless it serves a specific purpose, certain operators are only available in a particular representation, or the intended target representation is different from the source representation.

Conversion from Raster Data to Vector Data

Smoothing of the outlines of complex patches which may require conversion from raster to the vector structure is a typical example. The visual quality that can be achieved by smoothing algorithms in the raster domain is limited by the resolution of the dataset; hence, as shown in Fig. 6.27, some other raster-based methods may be selected to work with resampling of the raster grid to a higher resolution (i.e., oversampling). By the oversampling factor used (e.g., 4) each source raster cell may be divided into a number of sub-cells (e.g., 16 for factor 4). The smoothing effect is achieved by removing or adding sub-cells. This method can result in a considerable increase of the amount of data and stepped lines may still be visible unless the size of the sub-cells is below the minimum visual separability distance.

The main task of the smoothing operation is to remove stepped lines. As Peter (1997) has shown, the commonly used line simplification algorithms (e.g., Douglas Peucker) cannot yield the desired results and may even destroy the effects of the previous generalization operations. Thus, Herzog et al. (1983) proposed a simple yet effective algorithm for the simplification of boundaries extracted from raster data, which identifies regularly stepped portions of polygon outlines and replaces them by straight lines (shown in Fig. 6.28). This method can easily be modified to meet specific requirements. After its application, various line simplification and smoothing algorithms can be employed for further refinement.

Conversion from Vector Data to Raster Data

It is not easy to get an example that requires vector data be converted to raster structure for generalization without subsequent re-transformation (bidirectional conversion). A possible application might be that the target representation is raster and that vector data (e.g., line or point features) need to be integrated with an existing raster dataset. In such cases, integration should take place before any generalization process is executed. Since positional precision and semantic information are partly lost during the conversion process, it would not be of great use to generalize vector data prior to vector-raster transformation. Furthermore, the structure of the existing

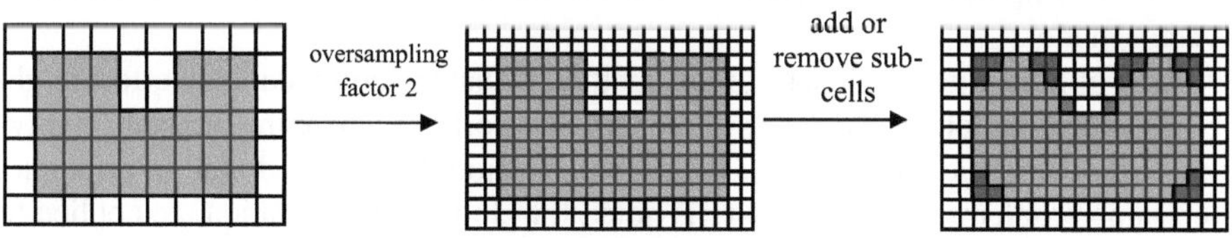

Fig. 6.27 Oversampling-based raster mode generalization

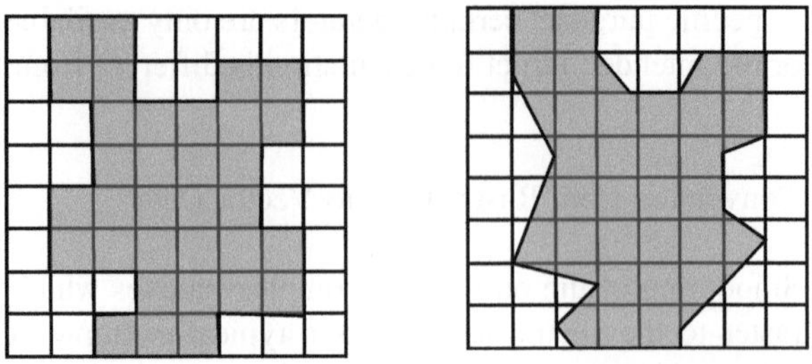

Fig. 6.28 Smoothing of raster boundaries by Herzog et al. (1983)

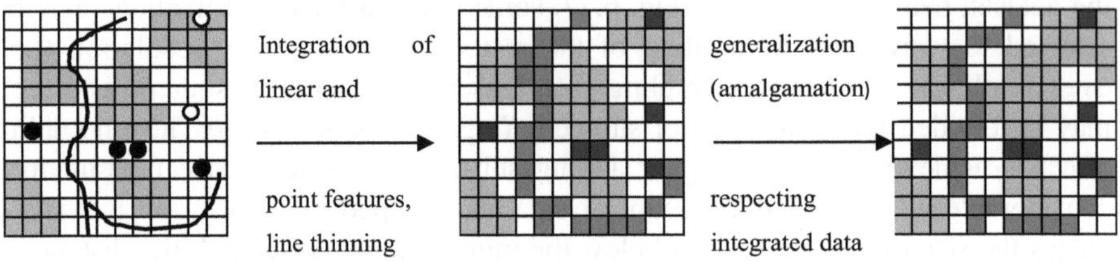

Fig. 6.29 Integration of vector data in the process of vector data to raster data conversion

regions will alter when data is integrated, as will the preconditions for the generalization process. Figure 6.29 illustrates the concept of converting and integrating vector data with raster data and subsequent raster-based generalization.

Local Conversion

A so-called local conversion method (Peter and Weibel 1999) can solve some of the problems mentioned in the above generalization strategies involving bi-directional data conversion. The principle of the local conversion is illustrated in Fig. 6.30. This conversion usually only occurs from vector to raster, where generalization operations are applied, followed by subsequent re-transformation to the vector model (i.e., vector-raster-vector).

One of the major advantages of the local conversion method is that the sampling resolution of the raster area may be coordinated with the specific properties of the patches involved and the planned generalization algorithms. In order to implement a local transformation, the minimum bounding rectangle of the desired polygons or area of interest can be calculated, and a margin may be added to avoid edge effects. After applying the generalization algorithms, the data should be re-converted to the vector model.

One of the major drawbacks of this method is that the computational effort might be considerable for large datasets and/or smaller portions where the methods involving local conversion are applied. In addition, it should be noticed that, in principle, the same problems and restrictions as noted for global transformation are applied to local transformations as well. Nevertheless, it is fortunate that since the amount of

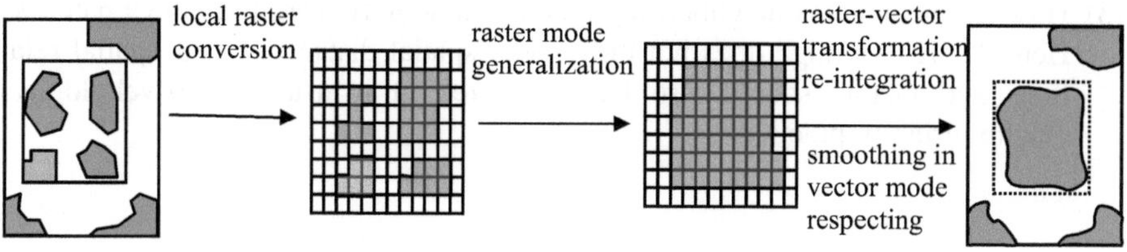

Fig. 6.30 Generalization based on bi-directional data conversion

data is generally rather small in most cases and all relevant parameters can be selected for a particular local situation, loss of information and positional precision can usually can be better controlled.

6.5 Summary of the Chapter

This chapter gives a detailed discussion of the description methods and generalization algorithms for continuous areal features on maps.

It firstly distinguishes discrete areal features and continuous areal features on maps; and then gives examples such as topographic maps and land use/cover maps to show continuous areal features. Secondly, it presents a number of key points or approaches for describing continuous areal features on maps, including topological relations among continuous polygons, topological relations on land-use maps, neighbouring relations on land-use maps, horizontal and vertical relations on land-use maps and geospatial semantic relations on land use maps. Thirdly, it discusses the principles and rules that should abide by in continuous areal feature generalization. After this, it addresses four types of algorithms/approaches for continuous areal feature generalization, including the rule-based land cover map generalization approaches, the areal patches generalization approaches considering spatial relations, the agent-based approaches for areal patches generalization, and the vector and raster-based approach for categorical map generalization.

At least the following problems regarding continuous areal feature generalization are worth further investigating:

(1) It seems that semantic relations of areal patches on land use/cover maps change when map scale changes, and it is obvious that they have a quantitative relation. How can this quantitative relation be expressed using a mathematical formula? Settlement of this problem is significant for fully automation of categorical map generalization.
(2) How to quantitatively evaluate the quality of the generalization of results of continuous areal features? This problem is almost the same as "how to evaluate the reasonability of a map generalization algorithm?"

(3) How to calculate the neighbouring relations among two arbitrary areal patches? Here, "neighbouring" does not only mean spatial distance. Other spatial relations and factors should also be considered. it is usually a psychological neighbouring in practice.

References

Adams B., McKenzie G., 2013, Inferring thematic places from spatially referenced natural language descriptions. In Crowdsourcing geographic knowledge, pp. 201–221, Springer.

Agarwal P., 2005, Ontological considerations in GIScience. International Journal of Geographical Information Science, 19: 501–536.

Ahlqvist O., Varanka D., Fritz S., Janowicz K., 2017, Land Use and Land Cover Semantics: Principles, Best Practices, and Prospects, CRC Press.

Ai T., Liu Y., 2002, Aggregation and Fusion in Land-use data generalization, Geomatics and Information Science of Wuhan University, 27(5): 486–492

Alameh N., 2003, Chaining geographic information web services. Ieee Internet Computing, 7: 22–29.

Ashburner M., Ball C.A., Blake J.A., Botstein D., Butler H., Cherry J.M., Davis A.P., Dolinski K., Dwight S.S., Eppig J.T., 2000, Gene Ontology: tool for the unification of biology. Nature genetics, 25: 25–29.

Bader M, Weibel R.1997. Detecting and Resolving Size and Proximity Conflicts in the Generalization of Polygonal Maps, the Proceedings of the 18th International Cartographic Conference, Stockholm, Swiden, Vol. (3):1525–1532

Baglioni M., Masserotti M.V., Renso C., Spinsanti L., 2007, Building geospatial ontologies from geographical databases. In International Conference on GeoSpatial Sematics, pp.195–209, Springer.

Ballatore A., Adams B., 2015, Extracting Place Emotions from Travel Blogs. In Proceedings of AGILE, pp. 1–5.

Berners-Lee T., Hendler J. and Lissila O., 2001, The Semantic Web, Scientific american 284 (5): 28–37.

Bishr Y., 1998, Overcoming the semantic and other barriers to GIS interoperability. International Journal of Geographical Information Science, 12: 299–314.

Bizer C., Lehmann J., Kobilarov G., Auer S., Becker C., Cyganiak R., Hellmann S., 2009, DBpedia- A crystallization point for the Web of Data. Web Semantics: Science, Services and Agents on the World Wide Web, 7: 154–165.

Brodaric B., 2004, The design of GSC FieldLog: ontology-based software for computer aided geological field mapping. Computers & Geosciences, 30: 5–20.

Buitelaar P., Cimiano P., Magnini B., 2005, Ontology learning from text: An overview. Ontology learning from text: Methods, evaluation and applications, 123: 3–12.

Buscaldi D., Rosso P., 2008, Map-based vs. knowledge-based toponym disambiguation. In Proceedings of the 2nd international workshop on Geographic information retrieval, pp.19–22, ACM.

Buttenfield B.P. & McMaster, R. 1991, Map Generalization: Making Rules for Knowledge Representation, London: Longman

Couclelis H., 2009, Ontology, epistemology, teleology: triangulating geographic information science. In Research trends in geographic information science, pp.3–15, Springer.

Couclelis H., 2010, Ontologies of geographic information. International Journal of Geographical Information Science, 24: 1785–1809.

Cresswell T., 2014, Place: An Introduction: John Wiley & Sons.

Cruz I.F., Sunna W., Chaudhry A., 2004, Semi-automatic ontology alignment for geospatial data integration. In International Conference on Geographic Information Science, pp.51–66, Springer.

de Berg M., van Kreveld M., Schirra S., 1998, Topologically correct subdivision simplification using the bandwidth criterion, Cartography and Geographic Information Systems, 25 (4):243–257

Di Donato P., 2010, Geospatial semantics: a critical review. In International Conference on Computational Science and Its Applications, pp.528–544, Springer.

Egenhofer M.J., 2002, Toward the semantic geospatial web. In Proceedings of the 10th ACM international symposium on Advances in geographic information systems, pp. 1–4, ACM.

Egenhofer M.J., Mark D.M., 1995, Naive geography. In International Conference on Spatial Information Theory, pp. 1–15, Springer.

Fonseca F.T., Egenhofer M.J., Agouris P., Câmara G., 2002, Using ontologies for integrated geographic information systems. Transactions in GIS, 6:231–257.

Frank A.U., 1997, Spatial ontology: A geographical information point of view. In Spatial and temporal reasoning, pp.135–153, Springer

Frank A.U., 2001, Tiers of ontology and consistency constraints in geographical information systems. International Journal of Geographical Information Science, 15:667–678.

Frontiera P., Larson R., Radke J., 2008, A comparison of geometric approaches to assessing spatial similarity for GIR. International Journal of Geographical Information Science, 22: 337–360.

Galanda M., 2003, Automated polygon generalization in a multi agent system, Dissertation for Doctor's Degree, University of Zurich, Switzerland.

Gao W., Gong J., 2005, Research on knowledge-based thematic data generalization in GIS, Geomatics and Information Science of Wuhan University, 30(5): 400–404

Gao W., Gong J., Li Z., 2004,Thematic Knowledge for the Generalization of Land Use Data, Cartographic Journal, 41 (3) :245–252

Gates A.Q., Keller G.R., Salayandia L., Da Silva P.P., Salcedo F., 2007, The gravity data ontology: Laying the foundation for workflow-driven ontologies. In International Conference on GeoSpatial Sematics, pp.278–287, Springer.

Golledge R.G., 2002, The nature of geographic knowledge. Annals of the Association of American Geographers, 92: 1–14.

Goodchild M.F., Hill L.L., 2008, Introduction to digital gazetteer research. International Journal of Geographical Information Science, 22: 1039–1044.

Harvey F., Kuhn W., Pundt H., Bishr Y., Riedemann C., 1999, Semantic interoperability: A central issue for sharing geographic information. The Annals of Regional Science, 33: 213–232.

Heath T., Bizer C., 2011, Linked data: Evolving the web into a global data space. Synthesis lectures on the semantic web: theory and technology, 1: 1–136.

Herzog A., Brassel K., Bopp M. Dorigo G., 1983, Ein System zur Generalisierung von Grenzliniendateien Administrativer Einheiten der Schweiz. Geoprocessing Series, Vol.3, Geographic Institute, University of Zurich

Hess G.N., Iochpe C., Ferrara A., Castano S., 2007, Towards effective geographic ontology matching. In International Conference on GeoSpatial Sematics, pp.51–65, Springer.

Hitzler P., Krotzsch M., Rudolph S., 2009, Foundations of semantic web technologies: CRC Press.

Hu Y., Janowicz K., 2016, Enriching Top-down Geo-ontologies Using Bottom-up Knowledge Mined from Linked Data. Advancing Geographic Information Science: The Past and Next Twenty Years, pp.183.

Hu Y., McKenzie G., Janowicz K., Gao S., 2015, Mining Human-Place Interaction Patterns from Location-Based Social Networks to Enrich Place Categorization Systems. In Proceedings of the Workshop on Cognitive Engineering for Spatial Information Processes at COSIT 2015. Santa Fe.

Hu Y.J., 2017, Geospatial Semantics. In Bo Huang, Thomas J. Cova, and Ming-Hsiang Tsou et al. (Eds): Comprehensive Geographic Information Systems, Elsevier. Oxford, UK.

Jaakkola O., 1998, Multi-scale categorical databases with automatic generalization transformations based on map algebra, Cartography and Geographic Information Systems, 25(4):195–207
Janowicz K., 2012, Observation-driven geo-ontology engineering. Transactions in GIS, 16: 351–374.
Janowicz K., Scheider S., Pehle T., Hart G., 2012, Geospatial semantics and linked spatiotemporal data–Past, present, and future. Semantic Web, 3: 321–332.
Janowicz K., Scheider S., Adams B., 2013, A geo-semantics flyby. In Reasoning web. Semantic technologies for intelligent data access, pp. 230–250, Springer.
Jansen L.J.M., Groom G., Carrai G., 2008, Land-cover harmonisation and semantic similarity: some methodological issues, Journal of Land Use Science, 3: 131–160
Jones C.B., Purves R.S., 2008, Geographical information retrieval. International Journal of Geographical Information Science, 22: 219–228.
Jones R., Zhang W.V., Rey B., Jhala P., Stipp E., 2008, Geographic intention and modification in web search. International Journal of Geographical Information Science, 22: 229–246.
Kuhn W., 2005, Geospatial Semantics: Why, of What, and How? Journal on Data Semantics III, LNCS, 3534: 1–24.
Lemmens R., Wytzisk A., d By R., Granell C., Gould M., van Oosterom P., 2006, Integrating semantic and syntactic descriptions to chain geographic services. Ieee Internet Computing, 10: 42–52.
Liu Y L et al. 2003. Categorical Database Generalization, Geo-Spatial Information Science, 6(4), pp.1–9
Lutz M., Sprado J., Klien E., Schubert C., Christ I., 2009, Overcoming semantic heterogeneity in spatial data infrastructures. Computers & Geosciences, 35:739–752.
Maedche A., Staab S., 2004, Ontology learning. In Handbook on ontologies, pp.173–190, Springer.
Mallenby D., 2007, Grounding a Geographic Ontology on Geographic Data. In AAAI Spring Symposium: Logical Formalizations of Commonsense Reasoning, pp.101–106.
Mark D., Egenhofer M., Hirtle S., Smith B., 2000, UCGIS emerging research theme: Ontological foundations for geographic information science.
Muller J.C. and Wang Z, 1992, Area-Patch Generalization: A Competitive Approach, The Cartographic Journal, Vol. 29, pp. 137–144
Neun M & Steiniger S. 2005, Modelling Cartographic Relations for Categorical Maps, XXII International Cartographic Conference, A Coruña, Spain, 2005
Nwana H.S., 1996, Software Agents: An Overview, Knowledge Engineering Review, 21 (3): 205–244
Peter B., 1997, Ableitung von Generalisierten Bodennutzungskarten aus der Arealstatistik der Schweiz 1979/85, Master thesis, Geographic Institute, University of Zurich
Peter B., Weibel R., 1999, Using vector and raster-based techniques in categorical map generalization, in the proceedings of the third ICA Workshop on Progress in Automated Map Generalization, Ottawa, 12–14 August 1999, pp.1–14
Piwowar J.M., Ledrew E.F. Dudycha D.J., 1990, Integration of spatial data in vector and raster formats in a geographic information system environment. International Journal of Geographic Information Systems, 4(4):429–444
Pundt H., Bishr Y., 2002, Domain ontologies for data sharing–an example from environmental monitoring using field GIS. Computers & Geosciences, 28: 95–102.
Purves R.S., Clough P., Jones C.B., Arampatzis A., Bucher B., Finch D., Fu G., Joho H., Syed A. K., Vaid S., 2007, The design and implementation of SPIRIT: a spatially aware search engine for information retrieval on the Internet. International Journal of Geographical Information Science, 21:717–745.
Qian H., Wu F., 2004, A monitoring model used in automated map generalization, Journal of Surveying and Mapping Technology, 21(3): 211–214
Ruas A.,2001, Automating the generalization of geographical data: the age of maturity, in the Proceedings 20th International Cartographic Conference, Beijing, China, 1943–1953

Schermer B.W., 2007, Software agents, surveillance, and the right to privacy: A legislative framework for agent-enabled surveillance, 21(3): 205–244
Schuurman N., Leszczynski A., 2006, Ontology-Based Metadata. Transactions in GIS, 10, 709–726.
Sen S., 2007, Two types of hierarchies in geospatial ontologies. In International Conference on GeoSpatial Sematics, pp.1–19, Springer.
Shamsfard M., Barforoush A.A., 2004, Learning ontologies from natural language texts. International journal of human-computer studies, 60, 17–63.
Shankar M., Sorokine A., Bhaduri B., Resseguie D., Shekhar S., Yoo J.S., 2007, Spatio-temporal conceptual schema development for wide-area sensor networks. In International Conference on GeoSpatial Sematics, pp.160–176, Springer.
Shea S. & McMaster R. B. 1989, Cartographic Generalization in A Digital Environment: When and How to Generalize, Proceedings AutoCarto 9, pp.56–67
Shvaiko P., Euzenat J., 2005, A survey of schema-based matching approaches. In Journal on data semantics IV, pp.146–171, Springer.
Smith B., Mark D.M., 2001, Geographical categories: an ontological investigation. International Journal of Geographical Information Science, 15: 591–612.
Sunna W., Cruz I.F., 2007, Structure-based methods to enhance geospatial ontology alignment. In International Conference on GeoSpatial Sematics, pp.82–97, Springer.
Visser U., Stuckenschmidt H., Schuster G., Vögele T., 2002, Ontologies for geographic information processing. Computers & Geosciences, 28: 103–117.
Wang Y., Gong J., Wu X., 2007, Geospatial semantic interoperability based on ontology. Geo-spatial Information Science, 10: 204–207.
Weibel R., 1996, A typology of constraints to line simplification, in the Proceedings of the 7th International Symposium on Spatial Data Handling, Delft, The Netherlands, Vol.1:9A.1-9A.14.
Weibel R., Dutton G.H., 1998, Constraint-based automated map generalization, in the Proceedings of the 8th International Symposium on Spatial Data Handling, Vancouver, BC, pp. 214–244.
Winter S., Freksa C., 2012, Approaching the notion of place by contrast. Journal of Spatial Information Science, 2012: 31–50.
Wooldridge M.J., 2002, An Introduction to Multiagent Systems, New York: John Wiley & Sons.
Wooldridge M., Jennings N.R., 1995, Intelligent agents: theory and practice, Knowledge Engineering Review, 10 (2): 115–152
Yan H. Weibel R & Yang B. 2008, A multi-parameter approach to automated building grouping and generalization, Geoinformatica, 12 (1), pp.73–89
Yan H., Yang W., Chen Q., and Liang T., 2000, Automatic construction of topological polygons, Journal of Graph and Image, 5(7): 563–567.
Ye M., Shou D., Lee W.C., Yin P., Janowicz K., 2011, On the semantic annotation of places in location-based social networks. In Proceedings of the 17th ACM SIGKDD international conference on Knowledge discovery and data mining, pp.520–528, ACM.

[illegible]

Chapter 7
Algorithms for Discrete Areal Feature Generalization

7.1 Introduction to Discrete Areal Features on Maps

Discrete areal features refer to the features on maps that are symbolized using polygons and the same type of features are topologically separated. Such examples are common on large or intermediate scale maps, e.g. lakes, ponds, seas, islands, buildings/settlements, parks, squares, playgrounds etc. (Fig. 7.1).

It is apparent that discrete areal features are usually easy to measure and locate in the geographic space, and their edges or boundaries are readily defined in vector or raster data model. If a single discrete areal feature is recorded in vector data, it is can be expressed using a number of vertices whose first and the last vertices are the same one; if it is recorded in raster data, it consists of a number of connected pixels.

Discrete areal features can be classified into two categories according to their geometric shape on maps: regular-shaped ones and irregular-shaped ones. The former includes only buildings or settlement, while the latter includes lakes, ponds, islands etc. This chapter focuses on description and generalization of the former but ignore the latter, for the former is most commonly used on various types of maps.

7.2 Description of Groups of Buildings/Settlements

7.2.1 Gestalt Principles for Describing Groups of Buildings

It has been noted that Gestalt principles have been applied for the recognition of spatial distribution patterns for many years in both digital and manual generalization

---

The original version of this chapter was revised: citation and reference has been included for Cetinkaya et al. 2015. The correction to this chapter is available at https://doi.org/10.1007/978-981-13-3678-2_9

H. Yan, *Description Approaches and Automated Generalization Algorithms for Groups of Map Objects*, https://doi.org/10.1007/978-981-13-3678-2_7

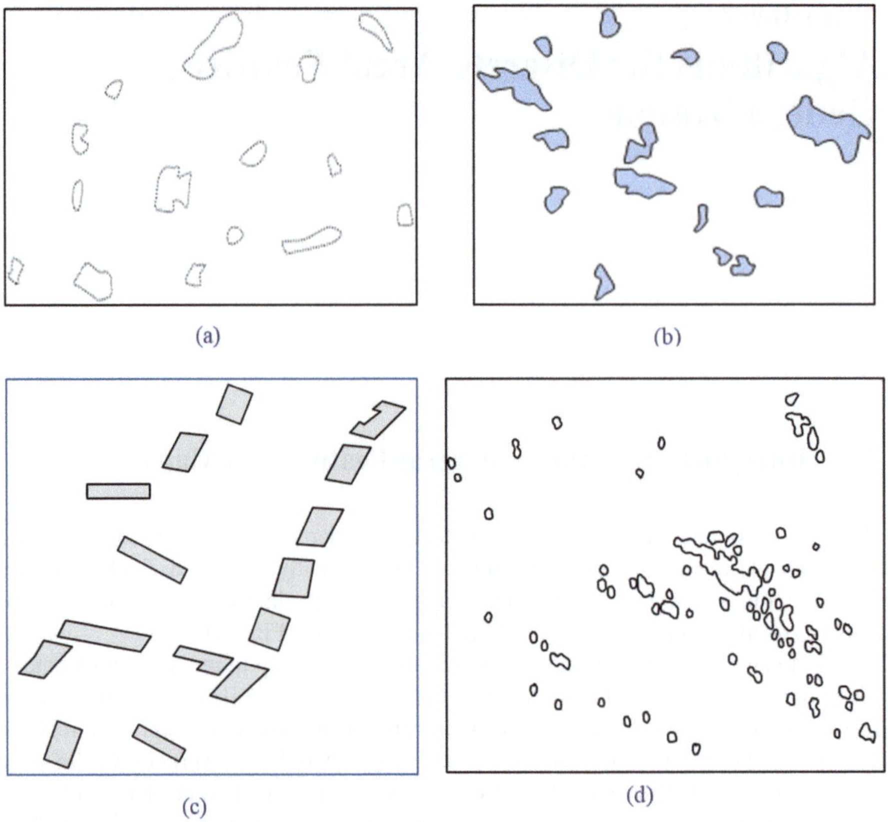

Fig. 7.1 Types of discrete areal features on maps. (a) vegetable land patches, (b) lakes, (c) settlements/buildings, and (d) islands

(Weibel 1996). In any case, from the point of view of perception, one conclusion from the research results remains unquestionable (Rock 1996):

> Grouping is not a simple, early process that works only on properties of image-based representations. Rather, it is a much more sophisticated process, one that incorporates the results of pictorial depth perception, occlusion, and amodal completion, as well as stereoscopic depth perception and lightness constancy.

Wertheimer (1923) took the initial step in identifying a number of important factors that have come to be known as the "Gestalt law of grouping", although "principle of grouping" is perhaps a more appropriate description. These principles are proximity, similarity, closure, continuity and common fate. To this list, two more factors have recently been added, i.e. common region (Palmer 1992) and element connectedness (Palmer and Rock 1994). For the detection of building alignments, the factor of common orientation plays an important role and should also be added to the list. To sum up, the Gestalt principles (Palmer 1992; Rock 1996) have so far been

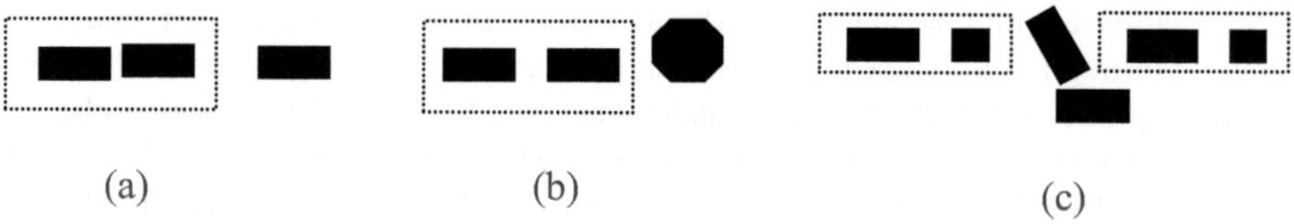

Fig. 7.2 Examples of rules for building grouping (buildings in dotted rectangles form a group). (**a**) Proximity: two close buildings form a group, while the distant one is separated. (**b**) Similarity: although the distances between pairs of buildings are equal, only the two buildings of same size and shape form a group. (**c**) Common direction: in case that distances, shapes and sizes are similar, only those objects that are arranged in the same directions form a group

employed in automated map generalization (Weibel 1996; Bader and Weibel 1997; Sadahiro 1997) may be presented as follows:

- Proximity: Objects at close distance tend to be perceived as a group (Fig. 7.2a).
- Similarity: Objects of similar shape and size perceptually form a group (Fig. 7.2b).
- Common orientations/directions: Objects arranged in a similar direction are perceived as a group (Fig. 7.2c).
- Continuity: Regularities or tendencies persist and are not easily disturbed. For example, two crossed curves maintain their respective continuity.
- Connectedness: Connected elements can easily form a group.
- Closure: An object group with a closed tendency is easily regarded as being perceptually closed.
- Common fate: The objects that are moving in the same direction appear to be grouped together (It is used in dynamic maps).
- Common region: Objects in the same region are more easily grouped together.

As far as map buildings within the scope of a specific street block are concerned, continuity, connectedness and closure imply that distances are zero. Hence, they can be substituted by proximity. The principle of common region is context-related, meaning that buildings in super-blocks are partitioned by streets and rivers, and are beyond the scope of this research which is restricted to buildings contained in street blocks. The principle of common fate is only relevant in dynamic maps. Therefore, only the first three of the above principles, that is, proximity, similarity and common direction, are taken into consideration for building grouping in this research (as illustrated in Fig. 7.2).

7.2.2 Two Types of Constraints for Buildings Grouping: Global and Local

Two steps are generally involved in the cognitive interpretation of visual signals: a pre-attentive phase and an attentive phase (Yan et al. 2008). In the former, an unconscious attempt is made quickly to extract information from an image or graphic through a global search operation that involves active grouping. The latter is a local phase, in which attention is paid to specific features of the visual landscape that have been identified as being different during the pre-attentive phase. That is, the global searching for groupings and locally focused identification are involved.

The visual information processing in map recognition is also similar to the above mechanism. It can be naturally obtained that cartographers usually also follow such a two-step process in the generalization of building features. In other words, there is a group of global constraints and a group of local constraints for the grouping and then the generalization of buildings.

7.2.2.1 Global Constraints

Buildings in cites are obviously more complex and therefore difficult to represent and generalize compared with that in country sides, thus more attentions should be paid on the buildings in urban areas. Urban morphology which deals with structures and/or patterns of cities is a well-established discipline, and some of the principles used in urban morphology may be adopted for describing groups of buildings and guiding map generalization. Indeed, it provides some factors that can globally control and/or express groups of building in cities.

Urban morphology may give the hierarchical structure of a city. In a classic neighbourhood model (Patricios 2002) in urban morphology, a city is, in essence, a hierarchical one comprising four levels (shown in Fig. 7.3). They are

- Enclave,
- Block,
- Superblock, and
- Neighbourhood.

Here, the most fundamental component is an enclave consisting of 20 or so houses. Three or four enclaves lined together may form a block. The enclaves within the block are separated from one another by a pedestrian pathway. The blocks are arranged around the sides of a central parkway in such a manner as to enclose the open green spaces. The clustered blocks, together with the central parkway, comprise a superblock. Several adjacent superblocks together form a neighbourhood bounded by major roads or natural features. Each neighbourhood is a fundamental unit of the city.

The roads in the neighbourhood are arranged hierarchically. Major traffic roads border each neighbourhood; distributor roads surround each superblock; cul de sacs provide access to individual property lots.

Of course, it is a natural thought that such a hierarchical partitioning of a city can be used for the hierarchical representation of urban features, such as buildings and roads, on maps at different scales. The four levels of the hierarchy must correspond

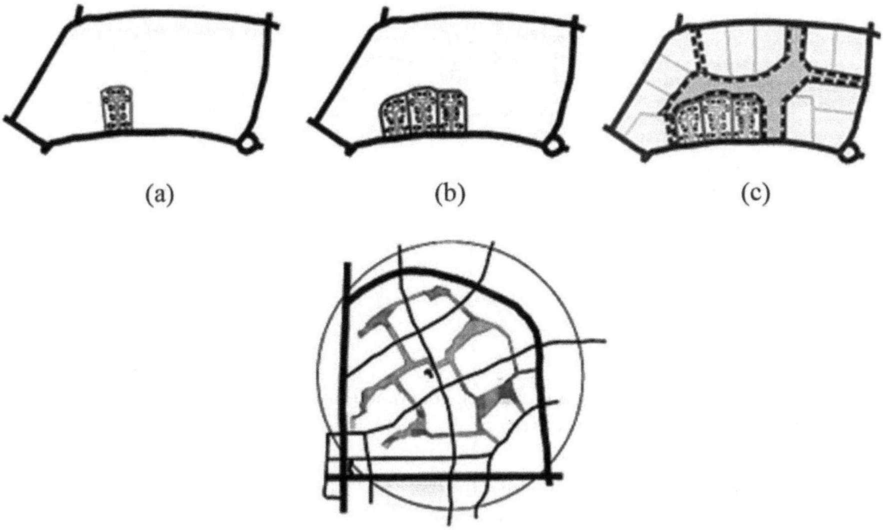
(a) (b) (c)
(d)

Fig. 7.3 Hierarchical representation of a city in Urban Morphology: (**a**) enclave, (**b**) block, (**c**) superblock, and neighbourhood

to a change in scale. Of course, if the change in scale is too great, then the whole city may become a point. But at such a change in scale, not much needs to be done in terms of generalization. Indeed, this neighbourhood model provides a macro view for the generalization of buildings with a limited range of change in scale and can be used to form global constraints.

The neighbourhood model plays two roles in the formation of global constraints for building grouping and generalization:

1. Neighbourhood model is used for global partitioning. In this step, an analysis f the whole map area is made. The features analysed are mainly roads and rivers, but not the buildings themselves. The axes of the roads and rivers are first extracted. The topological polygons of these crossed lines are then constructed. Each polygon stands for a partitioned global group of buildings. The resultant groups correspond to enclaves, blocks, superblocks or neighbourhoods due to different reductions in scale. Figure 7.4 is an example of global partitioning using the neighbourhood model. The extraction of medial-axes from streets is not a very sophisticated process, nor is the construction of topological polygons using crossed lines (Gold 1991; Klein and Meiser 1993; Christensen 1999). Therefore, no detailed method regarding this process is given here.

   What should be accentuated is that roads (and rivers if any) have a higher priority than buildings in the process of automated map generalization, as in the case of manual generalization. In other words, only after roads are generalized

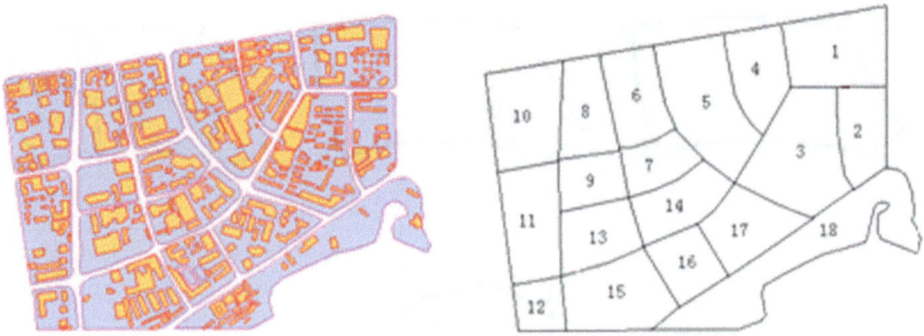

(a) (b)

Fig. 7.4 Global partitioning based on the neighbourhood model: (**a**) street map of a city, and (**b**) partitioning of the city by the street axes

can the global partitioning of buildings in the same area be calculated correctly. Thus, the building alignment is restricted to buildings within one street block.

2. After generalization, the topological polygons created based on the neighbourhood model are used as criteria to check if the relations among buildings and linear features (rivers and roads) have been well preserved, by which the harmony of the whole map can be ensured.

7.2.2.2 Local Constraints

Previous work (Boffet and Rocca 2001; Regnauld 2001; Li et al. 2004; Yan et al. 2008) suggests that the following parameters and their thresholds are particularly suited to describe buildings:

1. Major axis of a polygon: an axis is a line segment joining the mid-points of two edges of the polygon. The longest axis among them is defined as the major axis.
2. Minor axis of a polygon: this is an imaginary axis. It is perpendicular to the major axis and passes through the midpoint of the major axis. Its length is equal to the ratio between the area of the polygon and the length of the major axis.
3. Deviation angle: this is an acute or right angle formed by two crossed straight lines (they are generally major axes in building grouping).
4. Approximate parallels: if the deviation angle of two lines is less than 10° (experimental value), the two lines are regarded as being approximate parallels.
5. Approximately equal length: if $L_{max} - L_{min} < L_{limit}$or $(L_{max} - L_{min})/L_{max} < 0.5$, the two line segments are defined as being approximately equal in length.

   Here, L_{max} is the length of the longer line segment and L_{min} is the shorter one. Based on experience, the following thresholds are used for these parameters.

6. Length threshold, L_{limit}: the length of the shortest edges that can be represented on buildings, generally 0.3 mm.
7. Area threshold, A_{limit}: the area of the smallest building that can be represented on maps by its real size, generally $0.6 \times 0.4\ \text{mm}^2$.
8. Separation threshold, D_{limit}: the minimum distance between two buildings that is required for clarity, generally 0.5 mm.
9. Similar orientation: if one of the following two rules is satisfied, the two buildings are regarded as being of a similar orientation:
   - Major axes of the two buildings are approximately parallel; and
   - Major axis of a building and its minor axis are of approximately equal length; meanwhile, the major or minor axis of this building is approximately parallel with the major axis of another building.
10. Minimum distance
    It refers to the minimum distance between two buildings.
11. Area of visible scope
    It is a region between two objects. A detailed explanation of the definition of visible scopes is presented later in this section while discussing directional relations.
12. Similar size
    It can be represented by area ratio of the two buildings P and Q.

$$R_A^{P,Q} = \frac{A_{min}}{A_{max}} \tag{7.1}$$

Where, $R_A^{P,Q}$ is the area ratio of P and Q, and A_{max} is the area of the larger building, and A_{min} is the area of the smaller building.

13. Similar shape
    The edge number ratio may be used to evaluate shape similarity between two buildings P and Q.

$$R_E^{P,Q} = \frac{E_{min}}{E_{max}} \tag{7.2}$$

Where, $R_E^{P,Q}$ is the ratio of the number of edges of P and Q, and E_{min} is the number of edges of the building with fewer edges, and E_{max} is the number of edges of the building with more edges.

Using this function for the description of shape similarity between buildings is based on the fact that the internal angles of most buildings on maps are orthogonal. Hence, it can work well in most cases.

Fig. 7.5 Five measures for describing internal directions

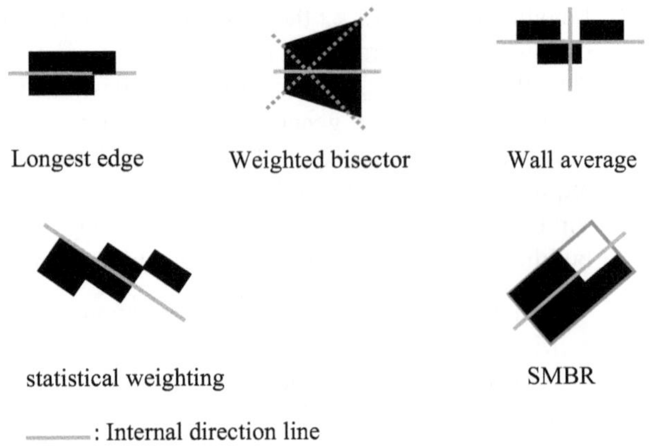

Fig. 7.6 Major axis and minor axis of an SMBR. Major axis: the longer axis of an SMBR. Minor axis: the shorter axis of an SMBR

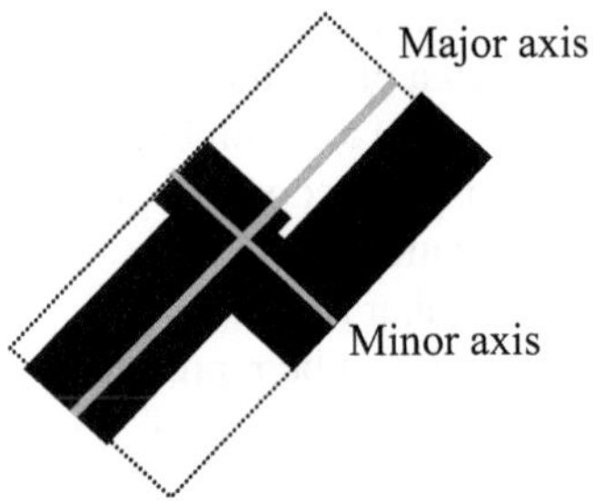

14. Internal orientation/direction

Internal orientation can be used to describe the spatial extent of an individual building. Duchêne et al. (2003) summarized five measures (Fig. 7.5), including longest edge, weighted bisector, wall average, statistical weighting, and smallest minimum bounding rectangle (SMBR), and concluded from their experiments that the SMBR is the most appropriate one.

To use the SMBR efficiently in building grouping, the major axis and minor axis are defined (as shown in Fig. 7.6) to denote the two main extents of a building. Major axes are more often used, however minor axes also may take an effect, especially if the directional extent of a building is not perceptually clear, that is, if the lengths of the major axis and minor axis are similar.

15. Directions between objects

The MBR model (Papadias and Sellis 1994; Shekhar et al. 1999), the cone model (Peuquet and Zhan 1987), the directional relation matrix model (Goyal 2000) and the direction Voronoi diagram (DVD) model (Yan et al. 2006) have been proposed for the description of direction relations. Here, the DVD model is employed because it can describe direction relations both qualitatively and quantitatively. The DVD

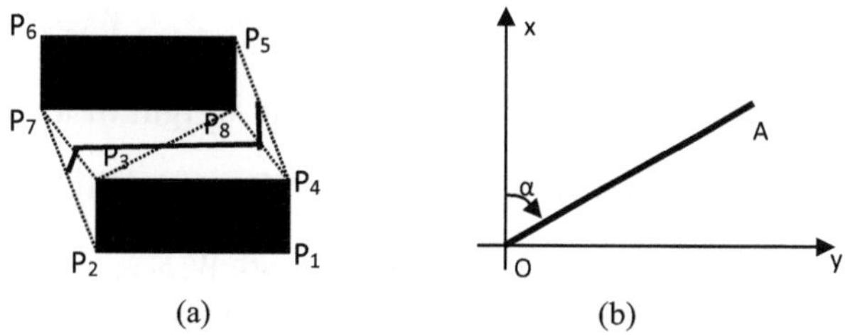

Fig. 7.7 Principles of the DVD model. (**a**) The thick line segments between the two objects are DVDs; (**b**) the definition of azimuths

model describes directional relations of two objects using DVDs. The procedure for the calculation of DVDs is as follows.

Step 1 Calculate the visible scope between two buildings: If a line segment connecting two vertices of buildings A and B has no intersection point (except the vertices themselves) with A and B and any other buildings, A and B are defined as 'visible to each other' and the two vertices are defined as 'visible vertices'. A scope between A and B whose boundary consists of all of the visible vertices is defined as 'visible scope' of the two buildings. For example, in Fig. 7.7a, the visible scope of the two buildings is the polygon $P_2P_3P_4P_5P_8P_7P_2$.

Step 2. Triangulate the visible scope: Take all vertices of the visible scope as a point set in two-dimensional space and construct the Delaunay triangles (Jones et al. 1995). Delete those triangles whose vertices belong to the same building.

Step 3. Compute DVDs: A curve may be formed by connecting the mid-points of each edge whose end-points belong to different buildings. The curve generally consists of several line segments. The normal line of each line segment denotes a direction, and the ratio of the length of the line segment with the total length of the curve is the weight (see a_1,a_2,. . .,a_8 in Formula (7.3)) of this direction.

All normal lines and their corresponding weights form a direction group, which presents a quantitative description of directions of two buildings. For example, in Fig. 7.7a, the DVDs are composed of three line segments which means three directions are needed for the description of direction relations of the two buildings.

Step 4: formation of directional group: In the eight-direction system (Goyal 2000), north means an azimuth (see Fig. 7.7b for the definition of azimuth) falling in (337.5°, 22.5°]; northeast means an azimuth falling in (22.5°, 67.5°]; east means an azimuth falling in (67.5°, 112.5°]; southeast means an azimuth falling in (112.5°, 157.5°]; south means an azimuth falling in (157.5^{o}, 202.5^{o}]; southwest means an azimuth falling in (202.5°, 247.5°]; west means an azimuth falling in (247.5°, 292.5°]; northwest means an azimuth falling in (292.5°, 337.5°]. While putting each azimuth into a cardinal direction, the corresponding weights belonging to the same cardinal direction are also added up, so that a quantitative description of directional relations of two buildings (say, P and Q) can be expressed by Formula (7.3).

$$\text{Dir(P, Q)} = \{< N, a_1 >, < NW, a_2 >, < W, a_3 >, \ldots < NE, a_8 >\} \tag{7.3}$$

Where, a_i (i = 1, 2, ……, 8) is the weight of a corresponding direction and $\sum_{i=1}^{8} a_i = 1$; and N, NW, …, and NE are the eight cardinal directions.

If Dir(Q, O) = $\{<N, b_1>, <NW, b_2>, <W, b_3>, \ldots < NE, b_8>\}$, the common directions of buildings P Q, and O can be expressed by Formula (7.4).

$$\begin{aligned} &\text{Dir(P,Q)} \cap \text{Dir}(Q,\text{O}) \\ &= \{< N, a_1 \vee b_1 >, < NW, a_2 \vee b_2 >, < W, a_3 \vee b_3 >, \ldots < NE, a_8 \vee b_8 >\} \end{aligned} \tag{7.4}$$

To facilitate the discussion, $a_i^{P,Q,O} = a_i \vee b_i$ is used to denote the weight of a common cardinal direction of objects P Q, and O.

7.3 Principles/Rules in Discrete Areal Feature Generalization

Generalization of groups of buildings should abide by the following rules:

1. Geographic characteristics of the mapping area should be taken into consideration in the map generalization. For example, a mountainous rural area where there are only a few settlements should be considered differently from an urban area where the space is crowed with settlements. For the former, the settlements should be retained if possible; for the latter, the settlements can be delete when necessary.
2. The purpose of the target map should be taken into account in the process of map generalization. It is usually an important factor in settlement selection and aggregation which takes effect in determining the minimum size of the settlement that can retained on the target map and the minimum distance between settlements that can be aggregated.
3. The ratio of the source map scale and the target map scale is the most important factor in map generalization. It determines to what extent the settlements can be generalized.
4. The distribution characteristics such as patterns and structures should be preserved before and after settlement generalization.
5. In selective generalization, comparison relations among the settlement densities of different areas should be preserved before and after settlement generalization.
6. After settlement generalization, the topological relation among settlements should be correct.
7. After settlement generalization, the topological relations among settlement layer and other layers (e.g., river layer, road layer) should be correct.

7.4 Approaches to Generalizing Groups of Buildings/ Settlements

Map generalization is a procedure for solving spatial conflicts and congestions in the process of generating smaller scale maps from larger scale ones using various appropriate operations (e.g. selection, displacement, simplification of map objects) under definite conditions (e.g. map scale, purpose, etc.). The construction of spatial data infrastructures (SDI) in many nations and regions and the widespread use of geographic information in computers in the context of geographic information systems (GIS) have brought with them the demand for automation of map generalization (Jones and Ware 2005). Indeed, automated map generalization has become an indispensable component and a hot issue in cartography and the GIS community in recent years (McMaster and Shea 1992; Ruas and Plazanet 1996; Li et al. 2004).

Here presents an algorithm for buildings generalization proposed by Yan et al. (2008) which focuses on automated building generalization in city blocks, with a focus on the step of detecting and forming meaningful building groups, preceding the actual generalization phase. Contextual features, such as roads and rivers, are not considered in the approach. To date, several methods for building grouping and generalization have been proposed. The approach for object grouping proposed by Steinhauer et al. (2001) uses the adjacency of buildings in the Voronoi diagram, the distance between buildings, and their cardinality as criteria to form building groups. The method was designed as a generic procedure for the recognition of so-called abstract regions in cartographic maps, consisting of disjoint map objects of arbitrary type (and not just buildings). Hence, while the use of the Vononoi diagram for that purpose is elegant and while the procedure offers the potential for the detection of arbitrary object groups the approach also ignores the use of grouping criteria that are more specific and important to buildings, such as directional or size relations among buildings. Christophe and Ruas (2002) also present an approach to building grouping, yet one that focuses on the detection of a very special kind of groups, that is, buildings aligned in rows. Regnauld (2001) shows how such aligned groups of buildings can be generalized by so-called typification, reducing the number of buildings in the generalization process, while maintaining the general pattern of the alignment. Boffet and Rocca (2001) present measures that can be used to characterize urban blocks and thw objects contained therein (incl. buildings). Their main interest is on the urban block as a whole, with the aim of classifying different types of settlement patterns. Their main contribution is the proposal of a series of measures to characterize the 'free space' between buildings. Rainsford and Mackaness (2002) concentrate on the simplification of the shape of individual buildings by means of template matching. Common to all methods discussed above is that they tend to focus on particular aspects of building groups or individual buildings only. For instance, directional relations among buildings have rarely been taken into consideration, and if so then only for special cases (e.g. Christophe and Ruas 2002 for the detection of building alignments). For this reason, our intention is to propose a new, more comprehensive approach, integrating directional relations with several other parameters.

Buildings are generally symbolized as discretely distributed rectangular polygons on large-scale and intermediate-scale maps. With the reduction of map scale, the maps become crowded and illegible. Hence, many buildings need to be deleted, merged, collapsed, or simplified, and so on. According to observations in the literature (Steinhauer et al. 2001) and cartographers' experiences in manual generalization, cartographers usually divide buildings into groups before generalization and then perform different operations on different building groups. Further analyzing and decomposing this process, it can be found that trained cartographers generally perform the following three successive steps in the process of building generalization:

1. A description of patterns and relations of buildings appears in cartographers' mind (mental map);
2. Buildings are clustered into groups; and
3. Appropriate operations are selected to perform the actual generalization process.

Computer-aided map generalization is an imitation of cartographers' behaviors. Hence, for the purpose of automated building generalization, the following three questions naturally arise:

1. How can a good description of the patterns, structures and relations of buildings be achieved?
2. How can buildings be combined into groups?
3. How can building groups be matched to appropriate generalization operations.

This research aims at answering the three questions and places the emphasis on questions (1) and (2), while question (3) is only partially addressed.

To facilitate discussion, the procedure for separating buildings into groups is called 'building grouping' in this algorithm.

7.4.1 Possible Operations for Building Generalization

In a digital environment, the generalization process has been decomposed into many operations (McMaster and Shea 1992), such as simplification, aggregation, combinations, collapse, etc. In particular, considering building features only, the following seem to form a subset of operations, i.e. aggregation, collapse, displacement, exaggeration, selective omission, simplification and typification. The meanings of these operations are derived from the work by McMaster and Shea, with occasional modifications.

- Aggregation: to combine buildings separated by space.
- Collapse: to make the feature represented by a symbol with a lower dimension, e.g. to represent a city as a point feature on small-scale maps.

- Displacement: to move the building to a slightly different position, normally to solve the problem of conflict.
- Exaggeration: to enlarge a building with small size for representation at a smaller scale maps on which it should be too small to be represented.
- Selective omission (or elimination): to eliminate small and unimportant buildings.
- Simplification: to make the shape simpler.
- Typification: to represent buildings with a typical distribution pattern, e.g. to represent groups of buildings aligned in rows (e.g. five rows) using fewer (e.g. three) rows.

Displacement and exaggeration can in fact be considered as a post-generalization process at the stage of graphic presentation (Li and Su 1995). Collapse means to represent an area feature using a line (e.g. river) or a point (e.g. a city). This happens only when the change in scale is very dramatic and the target scale is very small. A discussion of collapse is beyond the scope of this study. Therefore, building groupings will only be related to such generalization operations as aggregation, simplification, elimination, and typification. (In this algorithm, no differentiation is made between amalgamation and aggregation because the procedures for both operations are virtually the same.)

7.4.2 *Building Grouping*

To cluster buildings of a given street block into groups for the purpose of generalization, an iterative approach is utilized to form small, potential building groups first, and then combine them into intermediate groups, and eventually obtain the final groups appropriate for the target scale. This strategy is employed in the approach; therefore, the following three processes are discussed in this section:

1. Topological adjacency (or proximity) relations are detected, so that all 2-building groups are obtained.
2. Intermediate groups are constructed according to spatial relations of 2-building groups.
3. The intermediate groups with common buildings are either separated or aggregated, so that the final groups are achieved.

7.4.2.1 Detection of Potential 2-Building Groups

This process consists of three steps.

1. Triangulation of the buildings: Construct the Delaunay triangulation (Jones et al. 1995) between buildings using all vertices of the two objects, forming three types of triangles (Fig. 7.8a). If the three points of a triangle belong to the same building

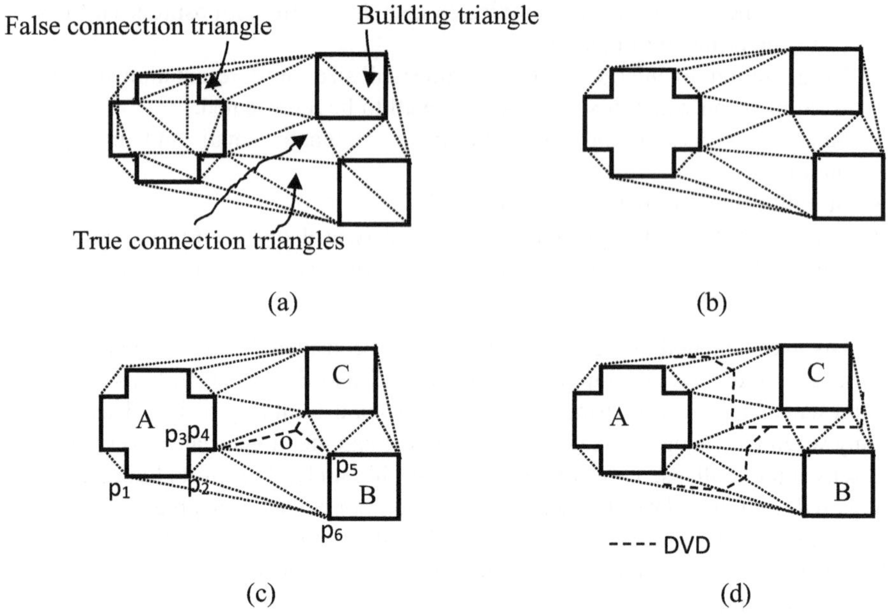

Fig. 7.8 Detection of topological adjacency relations and calculation of associated parameters. (**a**) Taking all vertices of the buildings as point set, the space is triangulated using the Delaunay triangulation rules; (**b**) building triangles are deleted from the triangle array; (**c**) visible scopes between every two adjacent buildings are constructed; and (**d**) the triangle array is traced and the DVDs of every two adjacent buildings are generated

and the triangle is located in the building, this triangle is called a 'building triangle'. 'True connection triangles' link two or three different buildings like a bridge. 'False connection triangles' are those that link the concave parts of a building. To detect topologically adjacent buildings, only the connection triangles are necessary. Hence, the building triangles are deleted from the triangle array, and only connection triangles are retained (Fig. 7.8b).

2. Detection of adjacency relations: In connection triangles, if three vertices of the triangle belong to the same building, this triangle is called a 'false connect triangle' (it is generally at a concave part of the polygon). Otherwise it is called a 'true connection triangle' (Fig. 7.8a). If a vertex of a true connection triangle belongs to a building, this triangle belongs to the building. Hence, every true connection triangle belongs to (and connects) two or three buildings. Two buildings owning a common triangle are defined as topologically adjacent, also implying a proximity relation.
3. Calculation of the parameters: The calculation of the parameters proposed in Sect. 2.2 is only carried out between two topologically adjacent buildings. Because it is easy to calculate area ratios, edge number ratios and SMBRs, we only discuss the calculation of the other three parameters here.

- Minimum distance: In the scope of every true connection triangle of two buildings, there must be a minimum distance between the two buildings. It is obvious that the minimum one of these distances in the scopes of true connection triangles is the minimum distance of the two buildings.
- Area of visible scope: The definition of the visible scope of two buildings has been discussed in the previous section of this chapter. However, the number of buildings is generally more than two in a block. In this case, there must be at least one triangle that belongs to three buildings. Hence, the problem of partitioning such connection triangles needs to be considered carefully. Here, for each such triangle, three line segments, each formed by connecting a vertex and the intersection point of the bisectors (i.e. the centroid), are constructed to partition the triangle. Take Fig. 7.8c as an example. The centroid of the triangle belonging to buildings A, B and C is o, so the visible scope of A and B is polygon $P_1P_2P_3P_4P_5P_6P_1$. A simple method for the calculation of the area of a visible scope is to sum up the areas of all triangles in the visible scope.
- DVD: In order to obtain the direction Voronoi diagram, the connection triangle array is searched to detect all true connection triangles. Using the methods introduced in the previous section of this chapter, DVDs between each pair of adjacent objects can be obtained (Fig. 7.8d). Then, the directions and their weights can be computed using the DVDs.

To save the above parameters, a structure named Parameter-Saver is defined as follows based on C++ (because the approach in this paper has been realized by the authors in C++), and a k x k matrix (say T) whose data type is Parameter-Saver is constructed to record the parameters of every two topologically adjacent buildings, where k is the number of buildings.

```
Typedef struct TagParameter-Saver{
    BOOL isAdjacent      //if two buildings are topologically adjacent
    float MinDistance;      //minimum distance
    float VisibleArea;      //area of visible scope
    float AreaRatio;      //area ratio
    float ENumRatio;      //edge number ratio
    float MajorAxis;      //major axis of SMBR
    float MinAxis;      //minor axis of SMBR
    float Drelation[8];      //weights of directional relations in
8-direction system
} Parameter-Saver
```

4. Formation of 2-building groups: Using the parameters recorded in matrix T, it is easy to obtain all potential 2-building groups, each of which is composed of two topologically adjacent buildings, denoted by the values of 'isAdjacent' in Parameter-Saver. An example for the construction of 2-building groups is shown in Fig. 7.9a and will be discussed further in the following section.

7.4.2.2 Construction of Intermediate Building Groups

To combine two 2-building groups, it is necessary to know:

1. The characteristics of the two groups themselves, and
2. The relations between the two groups.

The latter can be described using the parameters minimum distance and area of visible scope. For the former, a parameter named 'compactness' for evaluating whether a building group is 'strong', 'average' or 'weak' is defined. Here, three rules are employed, based on experience (Sadahiro 1997; Li et al. 2004) and the three Gestalt principles selected in the previous section of this chapter, to evaluate the compactness of a 2-building group (say, G):

1. If D > D_{limit} and A > A_{limit}, G is a weak group;
2. If D > D_{limit} and A $\leq A_{limit}$, or D $\leq D_{limit}$ and A > A_{limit}, G is an average group; and
3. If D < D_{limit} and A < A_{limit}, G is a strong group.

Here, the following definitions are used:

D_{limit}= 0.2 mm is separation threshold in map space (SSC 2005);
A_{limit}= 0.4 × 0.5 mm is area threshold in map space (SSC 2005);
D : minimum distance between two buildings; and.
A : area of visible scope between two buildings.

In general, for a group of n buildings (n $\geq$ 3), if D is substituted by $\bar{D}$ (mean minimum distance) and A is substituted by $\bar{A}$ (mean area of visible scopes), the above three rules can be also used to evaluate the compactness of intermediate groups.

Weak groups are subsequently deleted; while strong and average ones are retained (such an example is shown in Fig. 7.9b).

These retained groups cannot yet be directly generalized, for they are too segmented and have the potential to be aggregated to intermediate groups. Therefore, a further combination of these small groups is necessary. Suppose that two groups $G_1 = \{A_1, A_2, \ldots A_n, \mathrm{B}\}$ and $G_2 = \{\mathrm{B}, C_1, C_2, \ldots, C_m\}$ exist. Here, B is the common building of the two intermediate groups (see Fig. 7.9). If one of the following criteria is satisfied, G_1 and G_2 can be aggregated into a bigger group G:

1. Both G_1 and G_2 are strong groups.
2. In case one group is strong and the other is average, it should have $R_a^{A_n,B} > 0.6$ and $R_s^{A_n,B} > 0.6$ and $R_a^{B,C_1} > 0.6$ and $R_s^{B,C_1} > 0.6$, or $a_i^{A,B,C} \geq 40\%$ and both of the acute angles, respectively, formed by the major axes of A_n and B, B and C_1, are less than 15°.
3. In case both G_1 and G_2 are average groups, it should have $R_a^{A_n,B} > 0.6$ and $R_s^{A_n,B} > 0.6$ and $R_a^{B,C_1} > 0.6$ and $R_s^{B,C_1} > 0.6$, or $a_i^{A,B,C} \geq 40\%$ and both of the acute angles, respectively, formed by the major axes of A_n and B, B and C_1, are less than 15°.

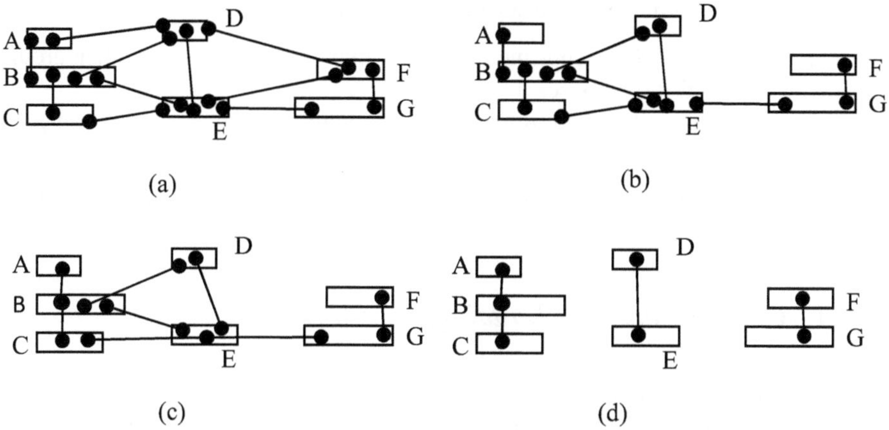

Fig. 7.9 Principles of building grouping. (**a**) Formation of 2-building groups (a dot-ended line segment connection denotes a 2-building group). (**b**) Deletion of weak groups (three connections between buildings *A* and *D*, *D* and *F*, *E* and *F* are removed). (**c**) Formation of intermediate groups, forming the six groups {*A*, *B*, *C*}, {*B*, *E*}, {*B*, *D*}, {*D*, *E*}, {*C*, *E*, *G*} and {*F*, *G*}. (**d**) Aggregation and separation of common buildings. Group {*A*, *B*, *C*} is stronger than groups {*B*, *D*} and {*B*, *E*}, and *D* and *E* belong to group {*D*, *E*} if it is separated from group {*A*, *B*, *C*}, thus building *B* only belongs to group {*A*, *B*, *C*}, that is, the connections between *B* and *D*, *B* and *E* are removed. Similarly, the connections between *C* and *E*, *E* and *G* are also removed and the buildings separated

At the beginning, G_1 and G_2 are 2-building groups; after the first round of aggregation, some become 3-building groups, so the compactness of each new group and D and A between each pair of potentially will-be-aggregated groups are re-calculated, and the above three rules are re-carried out to aggregate them. This process is repeated until no groups can be aggregated.

Figure 7.9b, c together illustrate an example of the construction of intermediate groups.

7.4.2.3 Formation of Final Groups

After the above procedure, the intermediate groups cannot be generalized yet, for some buildings appear in two or more groups simultaneously (see Fig. 7.9c). For the purpose of directly performing generalization operations on every group, the groups owning common buildings need to be either separated or aggregated. Suppose that G_1 and G2 are two groups and G_1 is stronger than G_2.

Here, the criterion for evaluating whether G_1 is stronger than G_2 is to compare the mean distance between buildings, the mean area of visible scopes between buildings, the mean value of common directions, the number of buildings, the mean value of area ratios, and the mean value of edge number ratios of the two groups in dictionary order. The greater the value is, the stronger the group.

The rules for aggregating and separating groups are:

1. If no building in G_2, except the common building, exists in any other group, merge G_1 and G_2; otherwise,
2. Delete the common building from G_2, i.e. the common building is only retained in the stronger group G_1.

After this step, the final groups that are ready to be generalized are achieved.

7.4.3 Generalization of Buildings

To generalize the building groups, appropriate generalization operations must be selected for each specific group. Let G be a group that will be generalized and let N be the number of buildings in G. The following rules are used to select operations for G:

1. If N = 1 and the area of the building is less than 0.4 × 0.5 mm in map space (SSC 2005), the operations are 'collapse' (to centroid) and 'enlargement'.
2. Otherwise, if N = 1 and the area of the building is not less than 0.4 × 0.5 mm in map space (SSC 2005), the operation is 'boundary simplification'.
3. Otherwise, if N = 2, the operations are 'aggregation' and 'boundary simplification'.
4. Otherwise, if N ≥ 3, find adjacent groups of G. If the characteristics of the adjacent groups are similar to that of G, the operations are 'typification' (here, 'typification' means 'aggregation' + 'division') and 'boundary simplification'. Alternatively, compare the free space (say, A_s) with the sum area of buildings (say, A_b) in this group. If $A_b > A_s$ the operations are 'aggregation' and 'boundary simplification'; otherwise, the operations are 'selection' and 'boundary simplification'.
5. Otherwise, the operations are 'aggregation' and 'boundary simplification'.

After all of the groups are generalized, the topological relations among generalized buildings need to be checked, since they may now overlap. For this purpose, generalized buildings are re-triangulated to detect their minimum distances. If two buildings overlap, a compromise method, slightly moving the buildings or slightly reducing the areas of buildings, is employed to settle this dilemma.

7.4.4 Experiments

The proposed approach has been implemented in Visual C++ (Version 6.0) and integrated into a topographic map generalization system named Automap. The functions and the criteria used in the approach are empirical, therefore various experiments have been carried out by the authors to test the approach. Three of them are shown in

Figs. 7.10, 7.11 and 7.12, respectively. Test data sets are provided by Shenzhen Municipal Bureau of Land Resource, Guangdong Province, China (see Figs. 7.10 and 7.11) and by the Institut Géographique National (IGN), France (see Fig. 7.12). Simulated data has also been used for particular purposes but is not shown here.

The three examples (illustrated in Figs. 7.10, 7.11 and 7.12) use topographic maps at the source scale of 1:10,000 to generate maps at scales 1:25,000 and 1:50,000, respectively. To demonstrate the adaptability of the approach, different types of data have been chosen, representative of different shapes and arrangements of buildings:

Experiment 1 (Fig. 7.10): Simple, mainly rectangular building shapes; parallelism between building groups but different orientations overall.

Experiment 2 (Fig. 7.11): Non-convex building shapes, but still mainly orthogonal in the corners; different orientations with little parallelism.

Experiment 3 (Fig. 7.12): Complex, non-convex building shapes with arbitrary corner angles; different orientations with little parallelism.

Delaunay triangulations and the separation lines of building groups (thick lines are used to separate groups) are depicted on the original maps, and the generalized maps are shown next to the original ones for comparison.

7.4.5 Discussions

The approach to automated building grouping and generalization answers the three research questions mentioned in the introductory section.

The first research question addressed the issue of achieving a quantitative description of relations and patterns of buildings. For this purpose, a number of parameters corresponding to three Gestalt principles, i.e. proximity, similarity and common direction, are selected. For proximity, these are the parameters minimum distance and area of visible scope; for similarity, these are the area ratio and edge number ratio; and for common direction, smallest minimum bounding rectangle (SMBR) and direction Voronoi diagram (DVD) are used.

The second research question dealt with building grouping. Here, Delaunay triangulation is used to detect topological adjacency relations of buildings and generated 2-building groups firstly, and then constructed larger, intermediate groups according to a set of rules. After aggregation and separation of intermediate groups owning common buildings, final groups were created which can be used as a basis for generalization.

Finally, the third question focused on matching the characteristics of the building groups generated in the previous steps to appropriate generalization operations and algorithms. A series of rules based on parameters such as the number of buildings involved, the size of buildings, the ratio of building area and free space, etc. as well as threshold values such as a separation threshold and an area threshold, were established to perform this complex task.

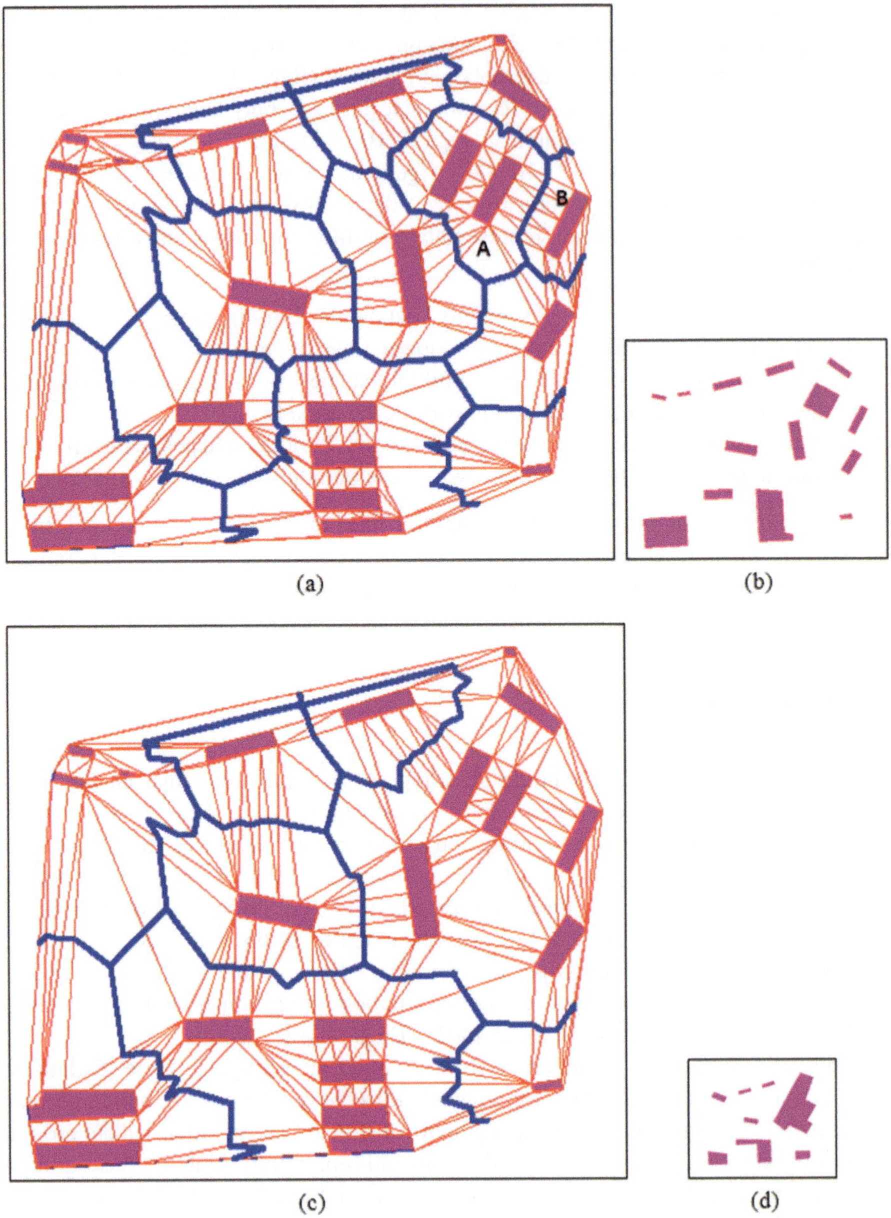

Fig. 7.10 Experiment 1 for building grouping and generalization (source map scale is 1:10,000): The buildings have simple and rectangular shapes, and have different orientations and much parallelism. (**a**) Grouping result for generating 1:25,000 map. (**b**) Generalized map at scale 1:25,000. (**c**) Grouping result for generating 1:50,000 map. (**d**) Generalized result at scale 1:50,000. All maps not shown to scale

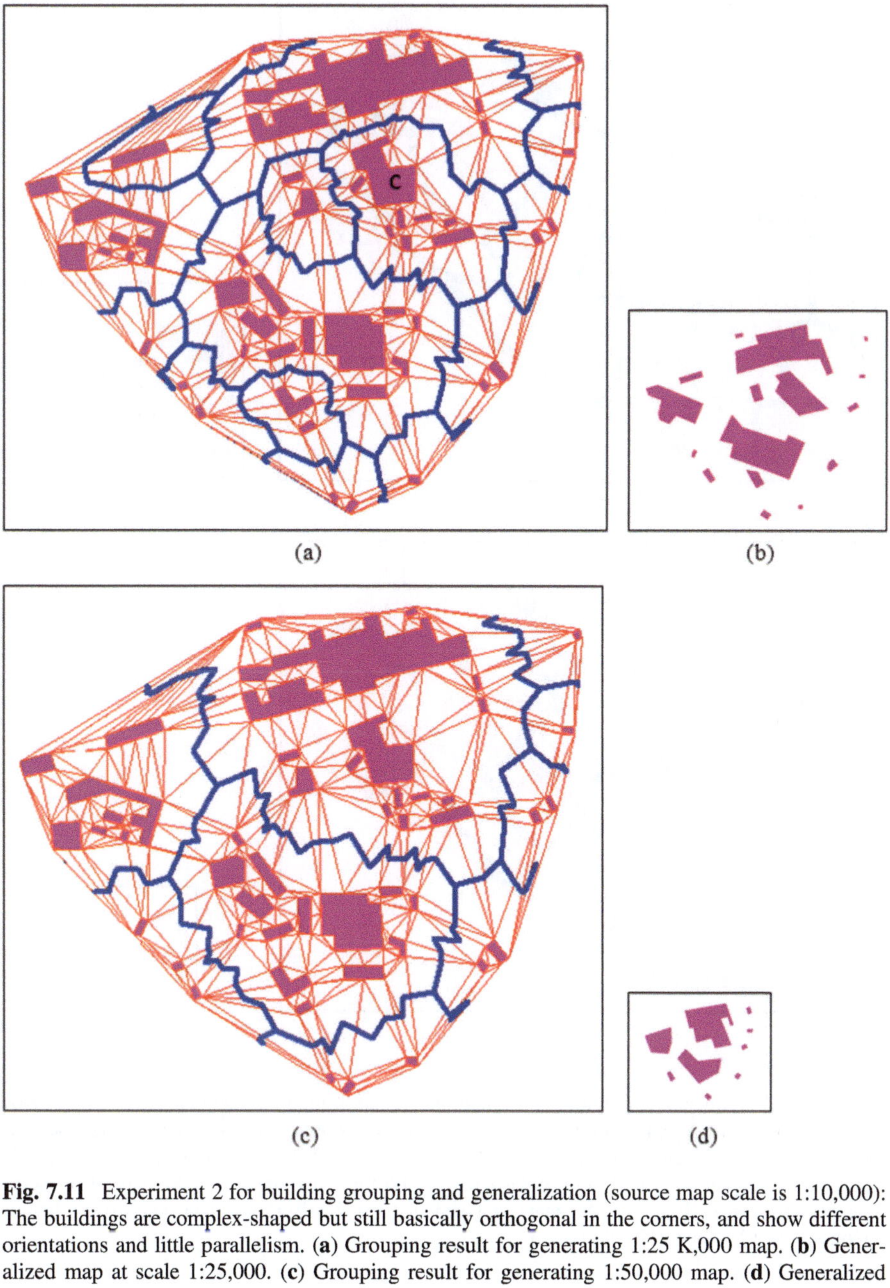

Fig. 7.11 Experiment 2 for building grouping and generalization (source map scale is 1:10,000): The buildings are complex-shaped but still basically orthogonal in the corners, and show different orientations and little parallelism. (**a**) Grouping result for generating 1:25 K,000 map. (**b**) Generalized map at scale 1:25,000. (**c**) Grouping result for generating 1:50,000 map. (**d**) Generalized result at scale 1:50,000. All maps not shown to scale

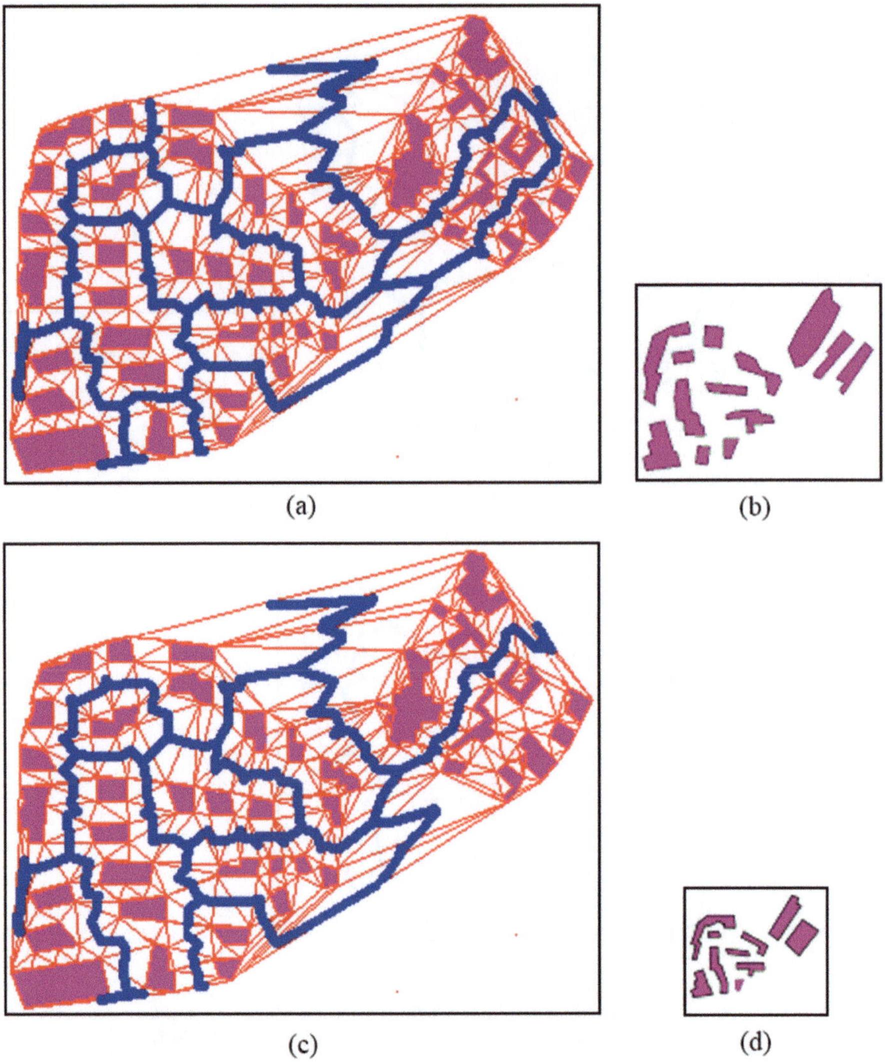

Fig. 7.12 Experiment 3 for building grouping and generalization (source map scale is 1:10,000): The buildings have complex and non-convex shapes with arbitrary angles in the corners, and have arbitrary orientations and little parallelism. (**a**) Grouping result for generating 1:25,000 map. (**b**) Generalized map at scale 1:25,000. (**c**) Grouping result for generating 1:50,000 map. (**d**) Generalized result at scale 1:50,000. All maps not shown to scale (BDTopo data courtesy of Institut Géographique National (IGN), France)

Further, a number of insights can be gained from the above algorithm and experiments.

Firstly, in building grouping, the principle of proximity takes priority over the other two principles. Hence, the parameters minimum distance and area of visible scope are prior to the other parameters. One such situation is shown in Fig. 7.10a: although three buildings in groups A and B are arranged in the same direction, the minimum distance and area of visible scope of the two buildings in group A are obviously less than that of the building in group B and the closest building in group A. Therefore, the two buildings with smaller minimum distance and smaller area of visible scope are clustered into one group. Group C in Fig. 7.11a shows a different kind of situation: the principles of similarity and of common direction almost can't work in this case, since the minimum distances are all less than the separation threshold and the areas of visible scope are all less than the area threshold, so they are directly clustered into a group based on the principle of proximity before the other two principles take effect.

The threshold values for D_{limit}=0.2 mm and A_{limit}=0.4 mm × 0.5 mm are from SSC (2005); while the other ones (e.g. $A_i^{A,B,C} = 40\%$ for common directions, $R_a^{P,Q} = 60\%$ for similar area, $R_s^{P,Q} = 60\%$ for similar size) used in the criteria for aggregating small building groups are based on previous experience and experiments (Li et al. 2004; Yan et al. 2006). They appear to work well in most cases, but occasionally lead to unsatisfactory results when the calculated values between two objects are just above or below their corresponding threshold values (traditional cartographers may face similar dilemmas when having to make discrete decisions in the generalization process).

In the experiments, each group that is formed for the generalization to smaller scale maps is exactly a combination of several groups used for the generalization of larger scale maps, though each grouping process is performed using the same source data set (i.e. each grouping process starts from the 1:10,000 scale data). This can easily be seen by comparing Fig. 7.10a with Figs. 7.10c, 7.11a with Figs. 7.11c, and 7.12a with Fig. 7.12c respectively. This is a kind of ideal grouping, as such hierarchical structures between different scales are traditionally also generated in manual generalization. Furthermore, there is potential to exploit these hierarchies in other tasks related to generalization, such as the calculation of the density of buildings over certain areas of a map at different aggregation levels.

Matching between building groups to generalization operators and algorithms (i.e. the third research question formulated in the previous section) is a considerably hard problem. This algorithm controls this process by means of a series of rules based on parameters such as the number of buildings, the area of buildings, the ratio between free space and sum area of buildings, etc. One of the operations not considered in the approach is map feature displacement. This could be applied as an optimization or post-processing step of the generalized results (Ruas 1998; Bader et al. 2005). In our work, conflicts are settled simply by slightly moving overlapping buildings or reducing their size. It is clear that more effort could be spent on addressing the third research question, that is, the matching of generalization operators to the characteristics of building groups. In the work presented here, however, it chose to focus primarily on the issue of building grouping, expressed by the two initial research questions posed in the previous section. Based on the building groups generated by our approach other, potentially better generalization algorithms could be used. Given the many parameters that are generated as a result of our approach, generalization algorithms will be well informed.

Finally, as the varied complexity and characteristics of the three test data sets seem to indicate, this approach has good adaptability in grouping and generalizing different types of buildings. It can be used to group and generalize buildings ranging from simple, rectangular shapes and regular arrangements, to non-convex, complex shapes in arbitrary directions and arrangements (e.g. Figs. 7.10, 7.11, and 7.12).

7.5 Potential Research Issues

Generalization of buildings or settlements is still an unsolved problem, needing further investigation, though many achievements have been made on it and fully automation generalization of buildings seems close at hand. The three questions put forward in the previous sections have not been satisfactorily answered, yet. On the other hand, some new concepts from other disciplines such as big data and cloud computing should have made map users proposed various initiative ideas and requirements that might bring map generalization or multi-scale representation of geo-spatial information into a new field and give us many new research issues. The following three topics may be listed among them, in the author's opinion, though there are many other ones:

- Multi-scale real-time maps

A main objective of traditional map generalization is to generate multi-scale geospatial databases. This type of generalization does not much care if the target map is real time; on the contrary, it pays attention to how to generate, organize and save multi-scale map data in geo-spatial databases.

Nevertheless, coming of big data makes generation of real-time, multiscale geospatial data theoretically. For example, trajectory data can be recorded by public transportation agencies as well as the private companies that do relative business. Such data can usually be collected and saved per day, per hour, or even per minute. If the trajectory data of a city is taken into account when the road network map of the city are generalized, the resulting map may be real-time and high quality.

Indeed, big data can play important roles in the generalization of many features on maps, especially those artificial features such as roads, buildings and vegetations, for people can record the change of these features conveniently. So, it should be a reasonable prediction that generation of real-time multi-scale artificial map features is one of the research issues in map generalization.

- High-speed generation of multi-scale geospatial data sets

Arrival of cloud computing provides a tool for us to produce multi-scale geospatial data sets at very high speed by means of parallel computing, which can be seen from the definitions and characteristics of cloud computing and parallel computing: on the one hand, cloud computing is an information technology paradigm that enables ubiquitous access to shared pools of configurable system resources and higher-level services that can be rapidly provisioned with minimal management effort, often over the Internet, relying on sharing of resources to achieve coherence and economies of scale, similar to a public utility; on the other hand, parallel computing is a type of computation in which many calculations or the execution of processes are carried out simultaneously, and large problems can often be divided into smaller ones, which can then be solved at the same time (Rauber and Rünger 2013; McCool et al. 2013).

If cloud computing and parallel computing are used in map generalization, map generalization not only can share computing resources provided by the Internet (i.e. by the cloud), but also may use parallel algorithms to improve map generalization efficiency by dividing the large map generalization task into small ones which can be solved by types of parallel algorithms. By this way, high-speed generation of multi-scale geospatial data sets can be achieved.

- Approaches to building grouping

Building groups can comprise different urban patterns based on similarity and regularity of geometric, semantic and structural characteristics of buildings and relationships between them (Cetinkaya et al. 2015). Maps portray these patterns with varying degrees of abstraction depending on scale and spatial context by means of generalization techniques. Hence, it is critical to detect these patterns at source scale and preserve them at target scales as much as possible so that maps can communicate geographic information without losing main messages throughout multiple scales. Moreover, generalization operators can produce better results in proximity-based smaller groups.

Grouping process is gradually performed in general because simultaneous use of all measures can produce too small and meaningless groups. Therefore, the first step of grouping is to analyze proximity relationships because close objects can be more spatially dependent or associated to each other. This idea can be supported by both geographic and cartographic foundations. From a geographic aspect, Tobler's first law of geography denotes: "everything is related to everything else, but near things are more related than distant things". From a cartographic aspect, individuals tend to visually perceive close objects in graphic representations as groups according to Gestalt principles (Cetinkaya et al. 2015). In addition, closely neighbouring objects have greater potential of graphic conflict (i.e. the violation of minimum distance constraint) at target scales in contextual cartographic generalization owing to enlargement and they should be treated as groups to solve this conflict (Yan et al. 2008).

Graph-based grouping methods constitute the most common approaches used in proximity-based grouping of buildings in urban blocks (Cetinkaya et al. 2015). Among them, Minimum Spanning Tree (MST) is the most widely used algorithm. Previous works related to building grouping for generalization usually utilize from auxiliary geometric data structures representing relationships. Regnauld (2001) generates groups of buildings through MST using centroids of buildings. In order to obtain subgroups of buildings that are subject to typification, the inconsistent edges to be eliminated are decided by proximity, homogeneity (i.e. orientation, size) and the number of buildings. Li et al. (2004) create Delaunay triangulation (DT) using the vertices of buildings. Buildings whose vertices belong to the same triangle are labelled as 'neighbour'. Direct alignments between neighbouring buildings are found by means of Gestalt theory. The indirect alignments among a row/column of buildings are then built upon the direct alignment relations. So, indirect alignments can be considered as special building groups. Anders (2003) demonstrates the success of neighbourhood graphs in finding object groups in a natural way without any parameter. Yan et al. (2008) use Delauney Triangulation to detect topological adjacency relations of buildings and generated 2-building groups firstly, and then constructed larger, intermediate groups according to a set of rules. After aggregation and separation of intermediate groups owning common buildings, final groups were created which can be used as a basis for generalization.

Although a number of approaches for building grouping has been developed, some questions such as "under what circumstances each of the approaches can be employed?", "how about the effectiveness of the approaches?" and "how about the quality of the results the approaches generate?" have not been well answered.

In addition, real-time big data may be the attribute data (some of them can also be spatial data) of corresponding map features and can be integrated in map generalization. This will increase difficulty in building grouping, and perhaps generate new research issues for us.

7.6 Summary of the Chapter

This chapter firstly gives an introduction of discrete areal features on maps, showing types of discrete areal features by examples. Secondly, it presents the approaches for describing groups of buildings or settlements on maps, including the Gestalt principles for describing groups of buildings and the two types of constraints for buildings grouping, i.e. global constraints and local constraints. Then it addresses a rules and principles that should obeyed in building generalization. After this, it discusses an algorithm for building grouping and generalization in detail, including its main ideas, the procedures and experimental results and some discussions. Lastly, it discusses three potential research issues on building grouping and generalization that can be considered in future work.

References

Bader M., Barrault M., Weibel R., 2005, Building displacement over a ductile truss, International Journal of Geographical Information Science, 19(8–9): 915–936

Bader M., Weibel R., 1997, Detecting and resolving size and proximity conflicts in the generalization of polygon maps, in the Proceedings of the 18th International Cartographic Conference, Stockholm, Sweden, pp.1525–1532

Boffet A., Rocca S., 2001, Identification of spatial structures within urban blocks for town characterization, in the Proceedings of the 20th International Cartographic Conference, Beijing, China, 2001 (CD-ROM)

Cetinkaya S., Basaraner M., Burghardt D., 2015, Proximity-based grouping of buildings in urban blocks: a comparison of four algorithms, Geocarto International, 30(6): 618–632

Christophe S., Ruas A., 2002, Detecting building alignments for generalisation purposes, In: DE Richardson, and P van Oosterom (eds.). Advances in Spatial Data Handling (10th International Symposium on Spatial Data Handling), Berlin: Springer-Verlag, pp.419–432

Duchêne C., Bard S., Barillot X., 2003, Quantitative and qualitative description of building orientation, in The 5th ICA workshop on progress in automated map generalization, Paris, France

Goyal R.K., 2000, Similarity assessment for cardinal directions between extended spatial objects, PhD thesis, The University of Maine

Jones C.B., Bundy G.L., Ware J.M., 1995, Map generalization with a triangulated data structure, Cartography and Geographic Information Systems, 22(4): 317–331

Jones C.B., Ware J.M., 2005, Map generalization in the Web age, International Journal of Geographical Information Science, 19 (8–9):859–870

Li Z., Yan H., Ai T., Chen J., 2004, Automated building generalization based on urban morphology and gestalt theory, International Journal of Geographical Information Science, 18(5):513–534.

McMaster R.B., Shea K.S., 1992, Generalization in digital cartography, Washington DC: Association of American Cartographers

Palmer S.E., 1992, Common region: a new principle of perceptual grouping, Cognitive Psychology, 24(2): 436–447, 1992.

Palmer S.E., Rock I., 1994, Rethinking perceptual organization: the role of uniform connectedness. Psychonomic Bulletin and Review, 1:515–519

Papadias D., Sellis T., 1994, The qualitative representation of spatial knowledge in two-dimensional space, Very Large Database Journal, 3(4): 479–516

Peuquet D., Zhan C.X., 1987, An algorithm to determine the directional relationship between arbitrarily-shaped polygons in the plane, Pattern Recognition, 20(1): 65–74

Rainsford D., Mackaness W., 2002, Template matching in support of generalization of rural buildings, In: DE Richardson, and P van Oosterom (eds.). Advances in Spatial Data Handling (10th International Symposium on Spatial Data Handling), Berlin: Springer-Verlag, pp. 137–151.

Regnauld N., 2001, Contextual building typification in automated map generalization, Algorithmica, 30(2): 312–333

Rock I., 1996, Indirect Perception. London: MIT Press

Ruas A., 1998, A method for building displacement in automated map generalization, International Journal of Geographical Information Science, 12(8):789–803

Ruas A., Plazanet C., 1996, Strategies for automated generalization, in the Proceedings of Spatial Data Handling, pp.6.1–6.18

Shekhar S, Liu X., Chawla S., 1999, An object model of direction and its application, Geoinformatica, 3(4):357–379

Steinhauer J.H., Wiese T., Freksa C., Barkowsky T., 2001, Recognition of abstract regions in cartographic maps, in: Montello DR, ed, Spatial Information Theory, Springer, Berlin, pp.306–321.
SSC, 2005, Topographic Maps: Map Graphics and Generalization, Cartographic Publication Series No. 17. Swiss Society of Cartography (CD-ROM)
Weibel R., 1996, A typology of constraints to line simplification, Advances on GIS II. M. J. Kraak and M. Molenaar (ed.), London: Taylor & Francis, pp. 9A.1–9A.14
Yan H., Chu Y., Li Z., Guo R., 2006, A quantitative description model for directional relations based on direction groups, Geoinformatica, 10(2):177–195
Sadahiro Y., 1997, Cluster perception in the distribution of point objects, Cartographica, Vol.34 (1):49–61
Wertheimer M., 1923, Law of organization in perceptual forms. In A Source Book of Gestalt Psychology, edited by W. D. Ellis (Kegan Paul, Trench, Trubner), pp.71–88
Yan H., Weibel R., Yang B., 2008, A multi-parameter approach to automated building grouping and generalization, Geoinformatica, 12:73–89
Patricios N.N., 2002, Urban design principles of the original neighbourhood concepts, Urban Morphology, 6(1), 21–32
Gold C.M., 1991, Problems with handling spatial data-the Voronoi Approach, CISM Journal, 45 (1): 65–80
Christensen, A.H.J., 1999, Cartographic line generalization with waterlines and medial axes. Cartography and Geographic Information Science, 26:19–32
Klein R.M., Meiser S., 1993, Randomized incremental construction of abstract Voronoi diagrams. International Journal of Computational Geometry and Applications, 3:157–184
Li Z.L., Su B., 1995, From phenomena to essence: envisioning the nature of digital map generalization, Cartographic Journal, 32:45–47
Rauber T., Rünger G., 2013, Parallel Programming: for Multicore and Cluster Systems. Springer Science & Business Media
McCool M., Reinders J., Robison A., 2013, Structured Parallel Programming: Patterns for Efficient Computation, Elsevier

Chapter 8
Concluding Words

This book systematically gives the description approaches and generalization algorithms of groups/clusters of features on 2-dimensional maps and summarizes latest and newest advancements in map generalization and multi-scale representation of geo-spatial information, for the purpose of laying a foundation for future research work and finding potential research issues.

Indeed, map generalization is an old and new topic in the community of cartography, which can be seen from its history and future: on one hand, it has thousands of years of history. We may even say that it came to the Earth along with the first map, because every map is a generalization of the Earth surface or a part of the Earth surface, though its definition has changed times and again with the change of mapping technology. On the other hand, modern information technology has thoroughly changed the map generalization techniques. It has thrown away manual map generalization and put map generalization into a kind of new medium, i.e. computer. Although rapid development of computer science and technology has made it come true that maps can be produced and transmitted via the Internet conveniently and quickly, it also has brought a number of new research issue regarding to map generalization. For example, users hope multi-scale maps can be generated within very short time (perhaps to make a map within several seconds) and the resulting multi-scale map data may be transmitted via the Web quickly and can be visualized at the user's end in a very short time. This actually requires researchers to explore online map generalization theories and techniques, by means of some other new techniques such as cloud computing and parallel computing.

In this information era, we should care about some upcoming research issues in map generalization beforehand so that we can prepare well when it really arrives. At least the following several points are worth our noticing.

- Big data and real-time multi-scale maps

  The emergence of big data presents a call to action for cartographers. As data sources creep closer towards the ability to describe every detail, all the time, for every place, the ways in which we produce multi-scale maps must adapt to handle

H. Yan, *Description Approaches and Automated Generalization Algorithms for Groups of Map Objects*, https://doi.org/10.1007/978-981-13-3678-2_8

this data windfall. New sources of information such as millions of trajectory data and conversations via location-enabled social media provides possibility to make real-time maps. These new data sources are of limited utility if we cannot find meaning in them; therefore, an overarching goal for cartographers is to find a way to put these data into the process of map generalization and make new, real-time, multi-scale maps that matter to people.

- Cloud computing, parallel computing and efficient map generalization techniques

  The purpose of map generalization should not only be for producing multi-scale geospatial databases that construct spatial data infrastructure for nations, regions and cities, but also be for producing online, real-time, multi-scale maps for various users. To achieve this goal a single computer generally cannot meet the basic requirements, and cloud computing and parallel computing which can share computing source online and provide high efficiency algorithms are feasible tools to use. Nevertheless, this research issue has not been deeply investigated, yet.

- Fully automated map generalization systems

  By far, fully automation of map generalization has not come true, yet, which can be seen from existing commercial software in map generalization. They involve an element of human control. Thus, it is emergent to systematically summarize existing algorithms, approaches and techniques in map generalization, by which we can find the gap between current situation and fully automated map generalization and naturally know what we shall do in the future.

Correction to: Algorithms for Discrete Areal Feature Generalization

Correction to:
Chapter 7 in: H. Yan, *Description Approaches and Automated Generalization Algorithms for Groups of Map Objects*, https://doi.org/10.1007/978-981-13-3678-2_7

The original version of this book was inadvertently published without citations to "Cetinkaya S., Basaraner M., Burghardt D., 2015, Proximity-based grouping of buildings in urban blocks: a comparison of four algorithms, Geocarto International, 30(6): 618–632".

The correct citation has been included as reference citation in text on page 223 and the reference has been listed on page 225.

The updated online version of this chapter can be found at
https://doi.org/10.1007/978-981-13-3678-2_7

H. Yan, *Description Approaches and Automated Generalization Algorithms for Groups of Map Objects*, https://doi.org/10.1007/978-981-13-3678-2_9

MIX
Papier aus verantwortungsvollen Quellen
Paper from responsible sources
FSC® C105338

If you have any concerns about our products,
you can contact us on
ProductSafety@springernature.com

In case Publisher is established outside the EU,
the EU authorized representative is:
Springer Nature Customer Service Center GmbH
Europaplatz 3, 69115 Heidelberg, Germany

Printed by Libri Plureos GmbH
in Hamburg, Germany